Do it yourself wooden garden constructions including wood and tool knowledge

AF289995

Preface:

I am a carpenter, trainer and writer. In this book you learn something about the material wood and the tools you have at home when you do home improvement work because you enjoy building something yourself.
Of course, this book can also awaken the love for wood crafts.
Our motto is: "Make your garden a personal paradise, I wish you a lot of fun."

Recommendations: Read the book up to the first construction projects, choose your next project from the instructions and read it through. Then read the project planning chapter. Follow these two guides, start your project and then dig deeper into the other guides if you have time or want to tackle a new one.
The animal enclosures are for garden animals except for the cat tree and the bird cage.

Note: Right-handed people attach the door hinges to the left and work with their right hand in the stable.
Left-handed people attach the hinges to the right (doors, windows... open to the right) and you can then work more easily with your left hand in the barn.
As a result, you have the door, the window... not on the side of your arm where you working hand with..

1. <u>Building material knowledge - wood</u>

When it comes to wood, there are two major distinctions. Softwood and hardwood can be both, conifer and deciduous tree.
Hardwoods are usually a bit more expensive to purchase, but they are not as susceptible to weather influences and/or pests or fungi' s.
You should remove the bark, because most insects or wood fungi romp in between wood and bark.
If you want to keep the bark, for aesthetic reasons, carefully separate the bark, paint the surface after you have rubbed it off vigorously. Brush the bark and dip it in lacquer for about 5 minutes so that the lacquer flows everywhere and seals. Then, after drying, glue the bark back onto the wood. Press the bark firmly so that each gap is closed. Then go with the paintbrush again over the connection points between bark and wood. By this way, you have preserved the aesthetics and still protected everything.
Most of the wood used in the constructions here comes from pine and spruce. This wood is very light in weight and workmanship. It is pale wood and can therefore be easily painted with all colors. Walnut, on the other hand, is an expensive hardwood that has a strong dark tint.
Of course, you can also opt for other woods because of the tint.
Hardwoods are heavier than the soft woods.

So You can build a garden bench from mahogany wood or from pine wood. If you paint the latter with mahogany-colored lacquer, you will hardly notice the difference except by the weight, if you paint it well.
Of course, hardwood is harder to work with. Birch can almost be scored with the fingernail, but is not suitable as a wood for tables or even the pavilion floor.

2. <u>Weather protection products</u>

There are various lacquers and stains. Choose the waterproof means for gardening. But beware, do not use any protective products in the "interior" of animal buildings, especially not for the ground. Birds scratch and peck on the ground and cats, hamsters, dogs and rabbits scratch on the ground. None of these products are suitable to be found in the stomachs... of the animals.
Think of the animals as your infants, who have to put everything in their mouths and maybe bite on them.
Stains and paints are usually not fireproof, (pay attention to the label).
Galvanized nails are also not recommended for animal constructions.
The tarring of dog- and other animal house roofing is highly dangerous. The tar (bitumen coating) becomes soft in the heat of the sun and glues the skin/feathers/fur of the animals. These then get movement difficulties and shortness of breath. They must then be taken care of by veterinarians.
Treat replacement parts on the outside first, allow them to dry and then replace the part.

3. <u>Tool knowledge and maintenance</u>

Saw blade for wood saws (fox tail and / or bow saw)

The toothing differs from that of the iron saw, except in the case of the jigsaw. This toothing is easier to clean of any resinification.

Chisels, planers and pull plane

Surfaces are smoothed and/or worked out with this. A pulling knife is used to remove the bark (peeling). When planing, peeling or punching, you always go with the stroke as with shaving. Go with the grain, otherwise you tear the wood open instead of removing small chips. Sometimes, wood can react more sensitive like a chin.

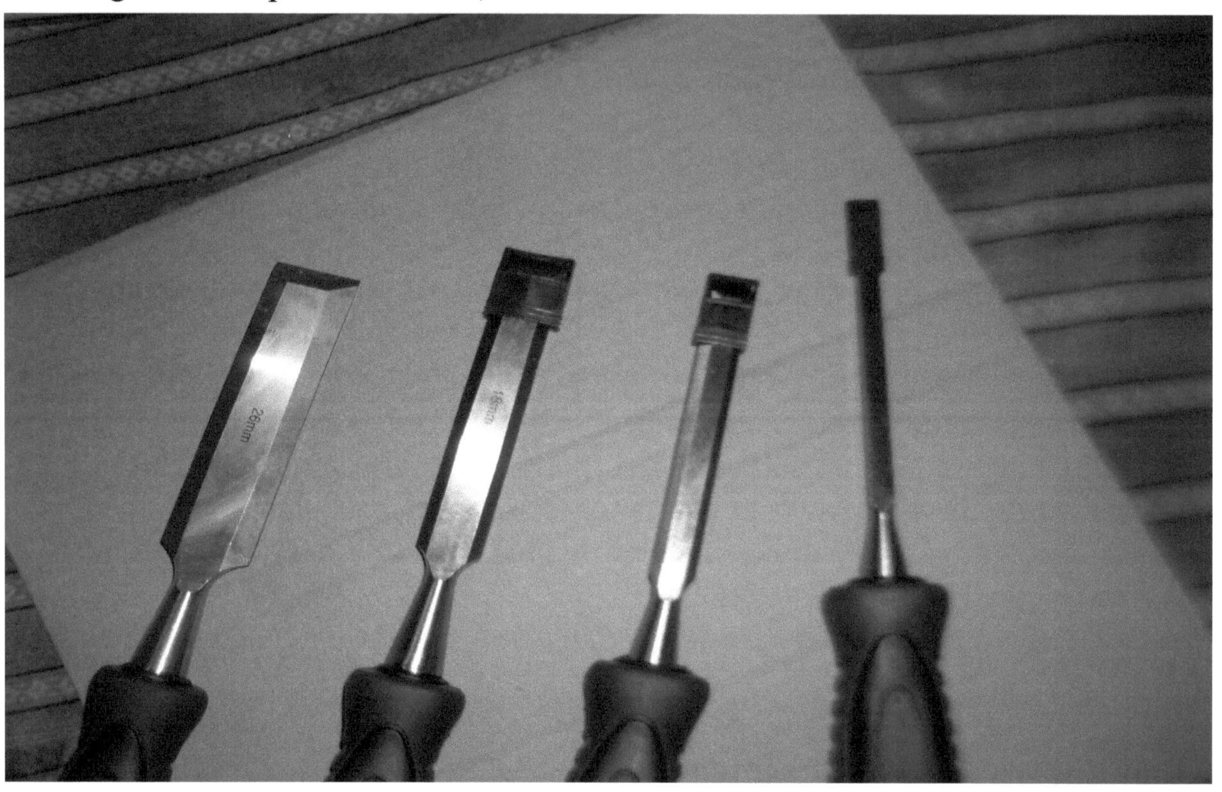

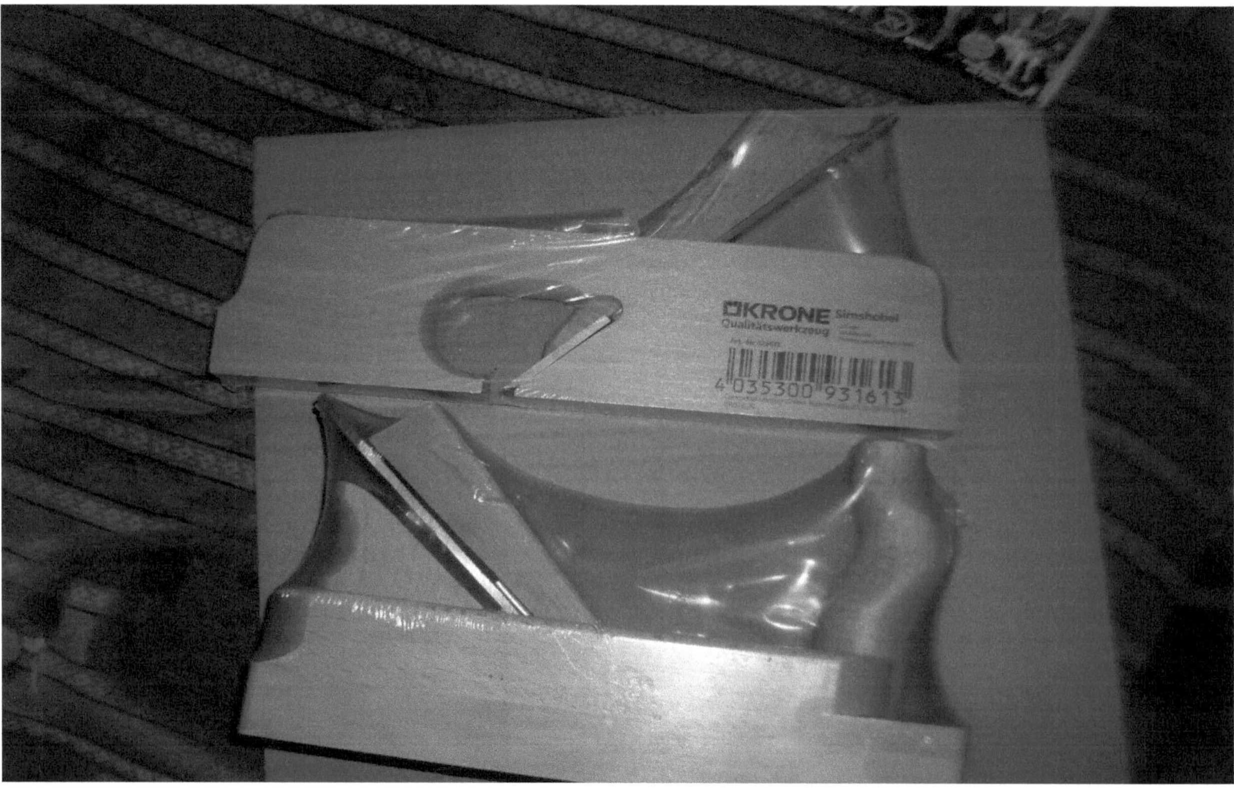

Rubber hammer/wooden hammer and carpenter's hammer
Rubber hammer or wooden hammer (mentioned as a rubber hammer hereafter), is
used for working with chisels so as not to damage them or, for example, joins like a
dovetail joint so as not to damage the surface of the wood.
The carpenter's hammer is suitable for hammering in nails, clamps and bolts.
If the nail has gone wrong, you can remove it again with the designated cow foot on
the hammer. The extended claw/tip of the hammer is used to prick under clamps, bent
nails or the like and lift them up until they can be removed with the entire cow foot.
The effort required for hardwood is much greater depending on the hardness. The
same applies to planing or punching out.
Note, working with hammer, is a work that also puts a lot of strain on the wrist. If
you are not used to this, it can lead to joint pain that recommends a break.

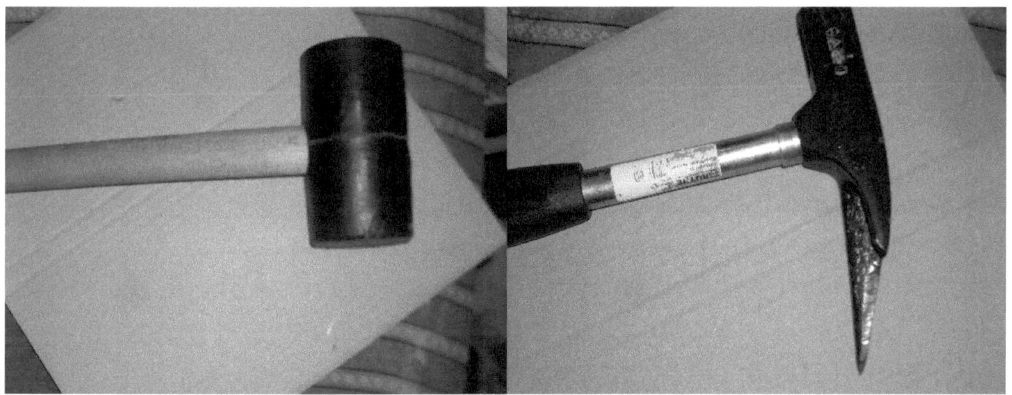

Angle iron, tape measure and solder, pencil, tear nail, spirit level

Angle irons are used to bring something into the right angle. The Pythagorean theorem is used to check whether this measuring instrument is not warped or bent in any way. To do this, mark on the short side at 3 cm, on the long side 4 cm on the outer edges. If you connect the two points with a tape measure and get 5 cm, the angle iron is still in order.

Tape measurements, whether made of metal or textile, are inaccurate if they have folds or notches. In the case of solder, the cord should not have a knot, but should be completely tightened by the weight of the solder body.

Carpenter's pencils are sharpened with a knife or chisel, as a woodworker does not carry an extra sharpener. For wood, tear nails are rarely or not at all used so as not to damage the structure of the wood. In the case of hardwood, it sometimes happens that a nail is used as a tear nail for marking similar to a pencil.

Spirit levels are used to see that you are properly level or like a solder to check that you are properly vertical.

How do you check or set a spirit level?

Place the spirit level on a table edge as if checking that the table is level (regardless of whether this is the case or not). If the bubble is exactly between the two lines (the table mid be horizontal). Now turn the spirit level 180 degrees remaining at the same edge of the table. If the bubble is again in the middle of the two lines, it means that the balance is in order and the table is also horizontal.

If the bubble is half over the line and after turning the spirit level is half over the line to the same extent as before (to the same side right or left as before turning) then the spirit level is fine, but the table is not horizontal, but is lower to the side where the bubble was drawn over the line. Otherwise, the spirit level must be straightened. In order to check the solder on the scale, you proceed as in the case of the table, only place the scale perpendicular to a corner of the room on the solder meter (the water bubble, which is located on the flat side of the scale), check again the level of the bubble located between the two lines, as in the case of the horizontal test. With most spirit levels, the solder meter cannot be corrected because the part is glued in.

To use a spirit level to check the vertical inspection on uneven walls, take a non-curved kitchen board and hold it against the wall, then place the spirit level on the board.

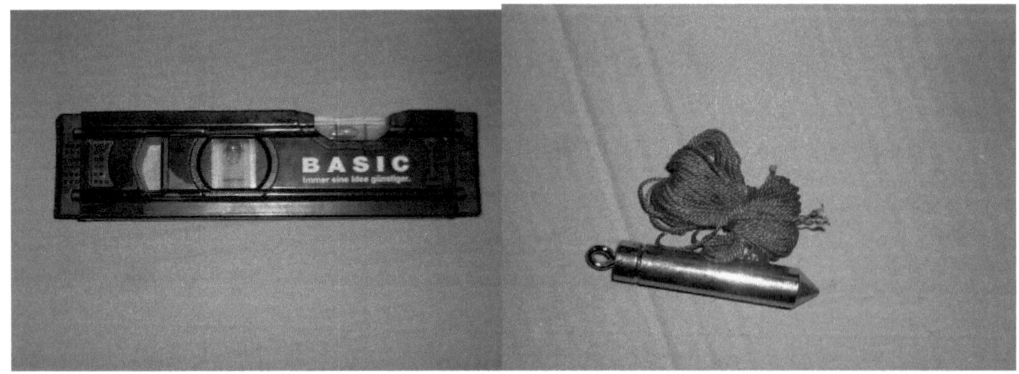

Spirit level   solder body with string

Brushes

Wide brushes are predominantly for smooth surfaces without joints or the like. Round brushes are for joints, inner edges or similar. This means that bristles from the flat brush do not bend into the joints and thereby fray.
For cleaning, it is important to pay attention to the label. Since most paints are supposed to be water resistant, water is not suitable for washing out and diluents must be used.

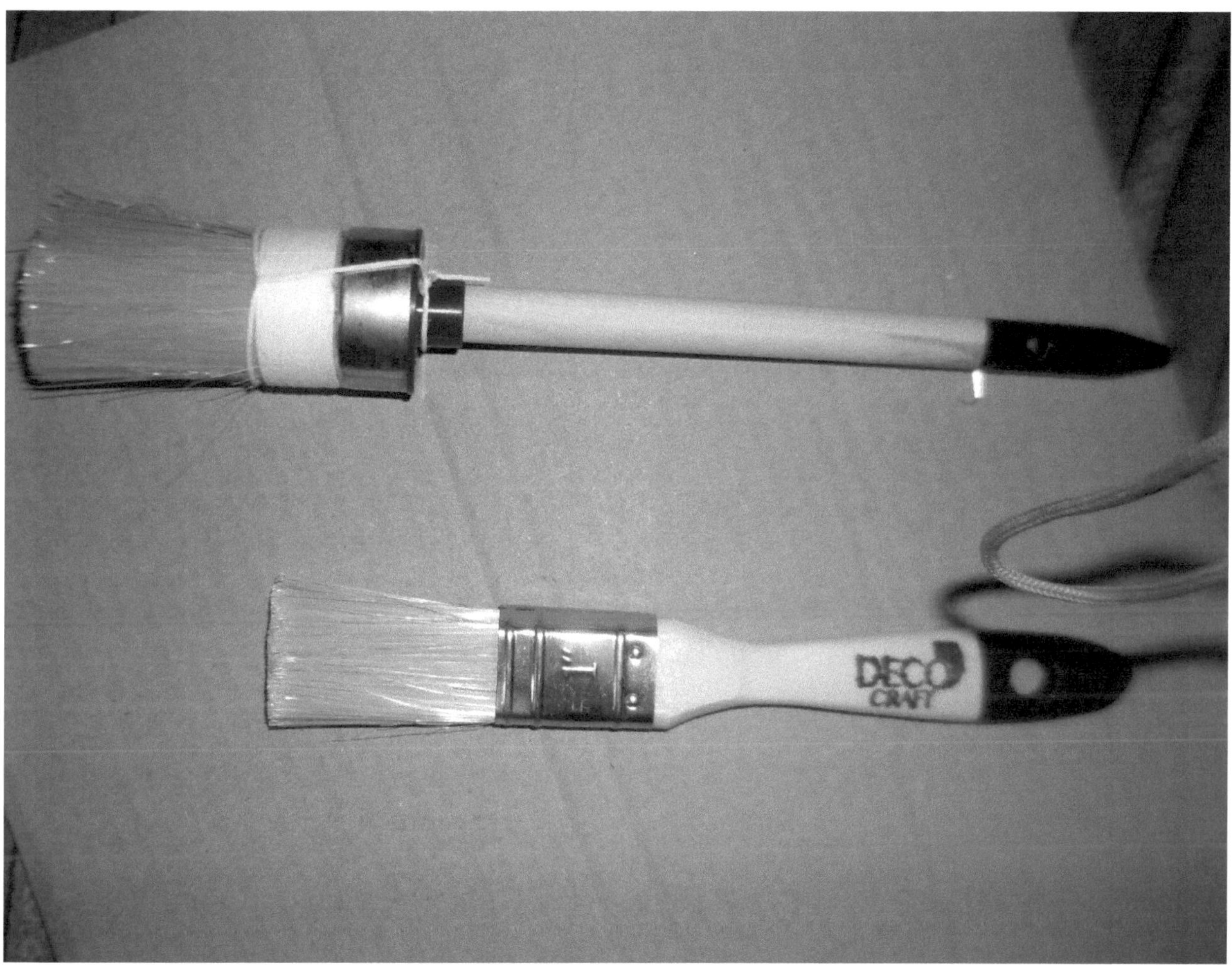

Wood drill bit

Drills differ in the hardening, tip and blade type that run along the drill core. In addition, you see drills here that are intended for drilling larger diameters. The toothing or knife edge can also be re-sharpened, as in the case of wood saws, by use of a knife or sword file.

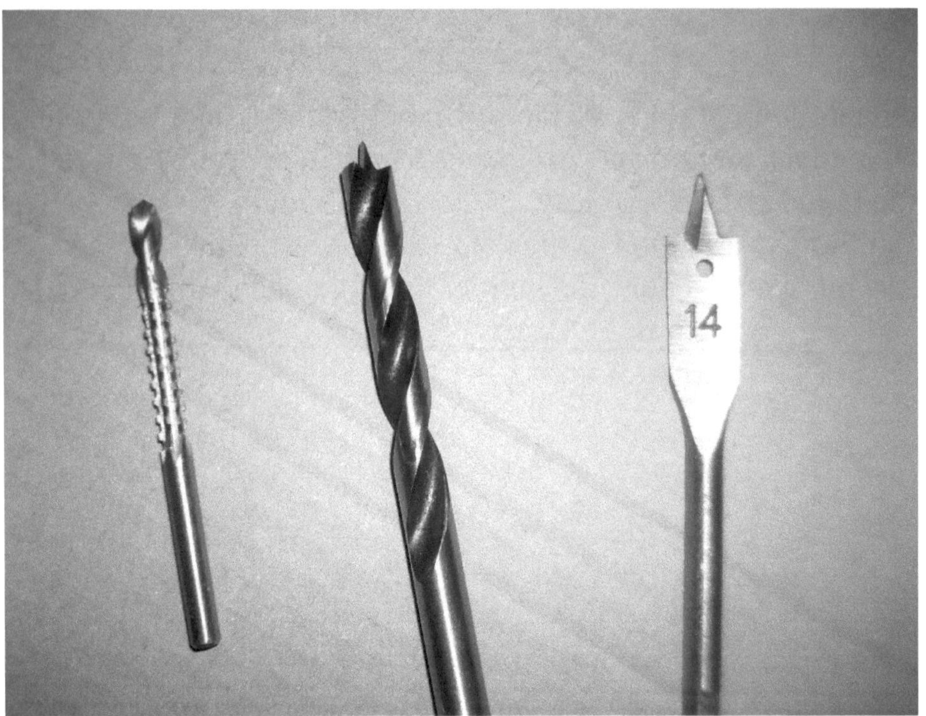

From left to right: wood sharpener, wood drill (clearly visible the protruding single tip), wood hole drill (here for 14mm hole diameter)

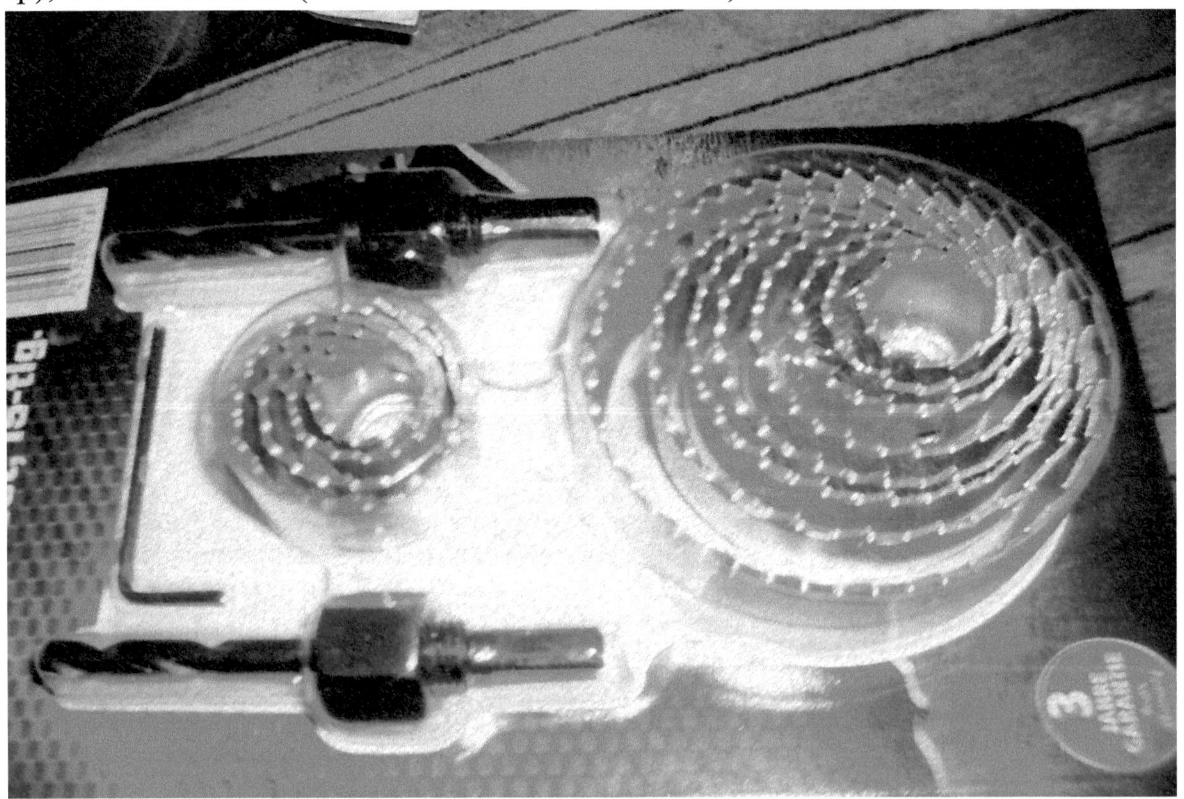

Hole Saws for Drilling Machines

Sword and/or knife file, sanding block, setting pliers

A sword file is shaped like a double-edged sword in profile, whereas the knife files are shaped like a knife in profile. Both are used for resharpening saw blades, their shape serves to reach between the tooth. It is only sharpened in the correct thrust, forward and **not** back again. Because otherwise the knife edge of the tooth will be

destroyed. So when you have finished with one side of the tooth, the saw blade has to be turned and the other side of the tooth is to sharpen. Chisels and planing knives are sharpened with a grinding block, namely only the oblique side of the knives. Dip the sanding block in water beforehand, this will increase the service life.
Thereafter smear the saw blades and files with a brush of oil. Briefly dip the knife in water and rub it with oil. When sharpening knives, slide the knife on the grinding block the inside (blunt side) to the outside and to the tip and not back. Also grind chisels, planing knives, hatches and axes only from the inside out and **not** back. Remove the scissors with the stone on the inside of the scissors towards the tip, **not** back. After some sharpening, it might be necessary to tighten the scissor screw. Clean free knives and scissors at least by rinsing off chips, even if you do not see any. Do NOT use abrasive paper, this usually tears and they penetrate into the table or other layers.

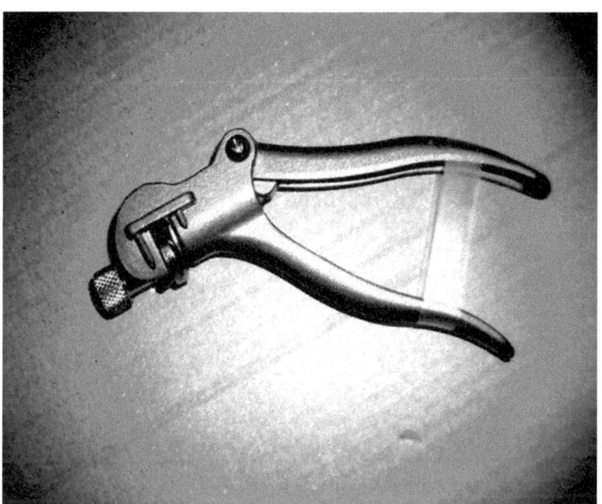

setting Pliers

Knife File

Knife file in profile

String:
With the string surfaces are measured in rolls/marked with reference to the Pythagorean theorem. A guide to this can be found, for example, in my book "Home Terra Preta – home made black soil"

4.) <u>Become one with the wood</u>

Get to know the wood you want to work with.

Wood is a living substance. Even if the tree has been cut long time ago, each board lives on for itself. By the way, many rootless plant parts can form new roots on their own under certain conditions.

There are two different types of wood/trees. These are hard and soft woods as well as deciduous and coniferous woods.

Whether a wood is soft or hard does not depend on whether it is a coniferous or deciduous tree. For example, the birch or pine is one of the softwoods, the birch is a deciduous tree and the pine is a coniferous tree.

Coniferous trees secrete resin more quickly. Although resin, the wound ointment of trees, is also an interesting raw material, it is extremely unpopular with wooden tools. Nevertheless, the very strongly resinous pine is often used as timber.

The hardness of the wood determines the amount of force required for machining and the degree of wear on the cutting tools. In nature, it increases the resistance to mold and insect infestation.

Precious woods are woods that have a special grain.

For example: black nut or mahogany

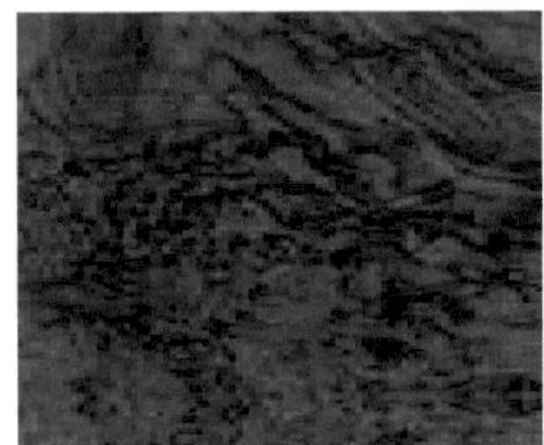  

Black walnut   mahogany wood   mahogany with high gloss lacquer

Such woods are not used for animal houses. They are more suitable as a bench or a table in the pavilion, although these woods are not only noble, but also have a higher price.

Branch scars can give the whole project an interesting look, but are less liked by saw blades and chisels because they are harder than the surrounding wood. It is advisable not to hit branch holes, but to cut these knife-like. This, of course, takes a little longer. Select the boards, lasts... preferably so that you do not have branch scars in places where you still want to do fine work.

Resin bags, (places filled with solid resin), can break out or fall out if they are solid. There remains a hull if they are not yet solid. Resin can stick to the machining tools. If a resin filling has fallen out, you can fill the area with wood glue, wood putty or lacquer.

Woodwork can go under the skin (splinters). A real home craftsman or even professional woodworker has tweezers in the home pharmacy or medical cabinet to pull wood splinters out of the skin again.

In order to properly process wood, you have to look at how it has grown. If you plan against the grain, you tear open the wood. The same is true of hitting with the chisel. On the photo of the painted mahogany wood, the grain can be best recognized. You always plan in the direction of "tongue tip". Planes or trunks towards them tear open this very "tongue". Handle the wood gently. The more you drive the knife in to hoist or plane, the greater the risk of tearing. Drive gently over the wood and feel the grain gradients. If they feel you hand would climb a "staircase" and not descend, you brush against the grain. Branches always grow upwards, with the grain.

5) <u>Timber truss connection types</u>

Timbered connections are among the oldest connections in the history of construction, in addition to wrapping by ropes or the like. They are intended to prevent slipping by drag, pressure or other movements and to ensure the stability of the connection and they were originally made only of wood itself.
Wood is thus connected lengthwise, used as a corner or cross connection.

Today, wood connections are additionally secured by glues and/or nails.
There are also steel stencils and others that are added in engineering, but in this book I do not go into more detail.

The blade butt joint is the simplest joint that can be used as an extension or as a corner joint. It resists compressive forces that act centrally in lengthen of both woods, but not to other separation movements.

Dovetail connection is a connection that is tension and pressure secured.
To the side, however, it can dissolve.

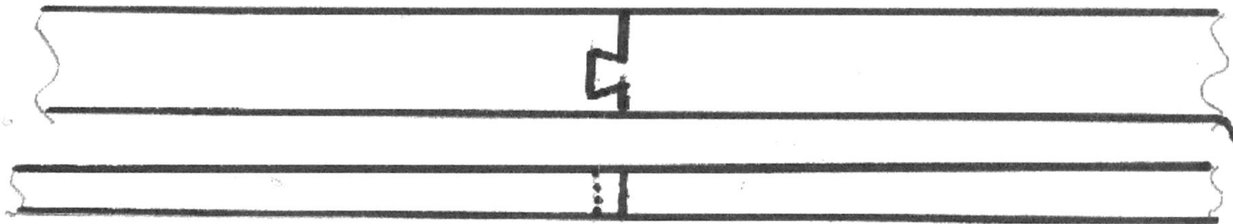

Due to the combination of dovetail and blade joint, dissolving to the side in one direction can be prevented.

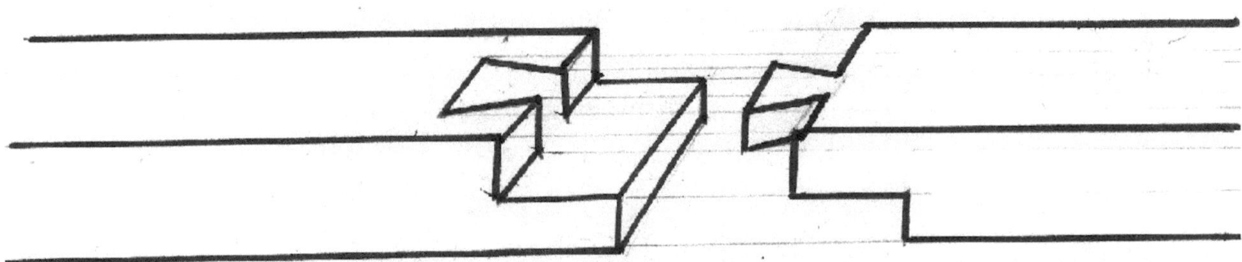

(Dovetail - Leaf Joint)

Spigot connection: instead of the dovetail, I can of course also make a straight spigot. Working up the length of the spigot or dovetail on a beam, this for example, also gives me the opportunity to use it as a guide rail.

A tenon-blade **joint** is like the dovetail blade joint

Hidden spigot is suitable for preventing the slipping of superimposed or standing woods.
A **spigot-(connection)** that is inserted into the other wood.

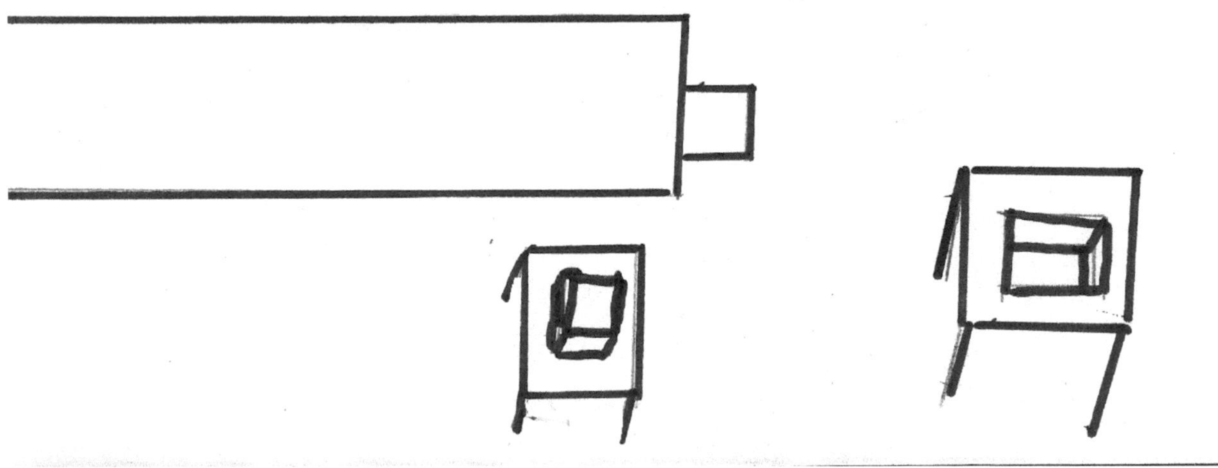

An **"oblique hook blade connection"** is usually used against the drifting apart of two parts. Therefore, it is used against compressive forces.

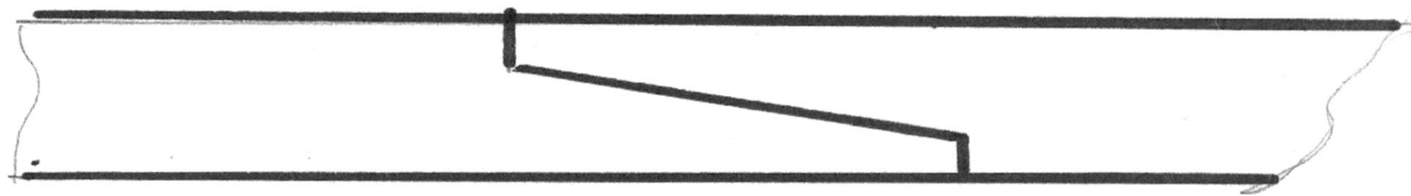

Hidden hook-leaf: "Look, I can conjure, it is not glued or nailed and still does not slip down."

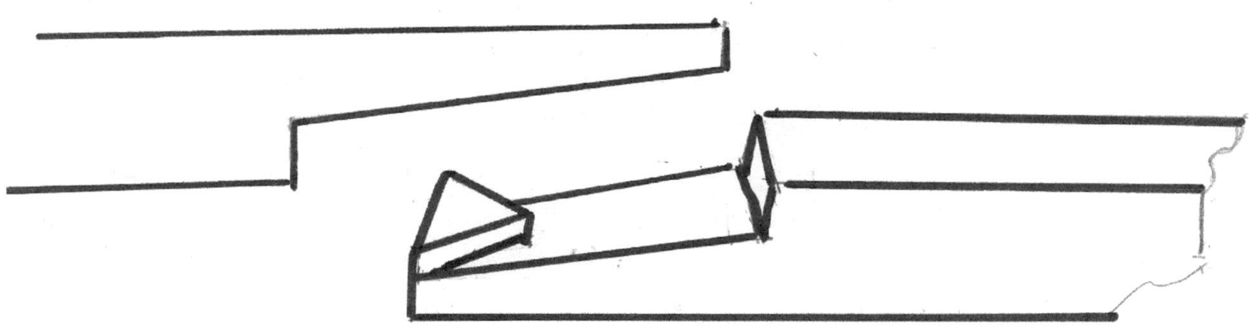

Pressure sheet connection:

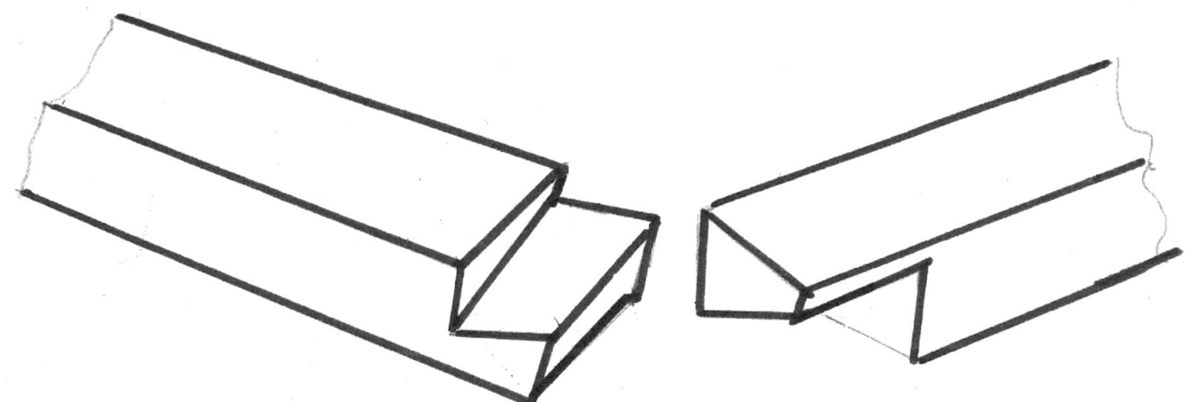

Cross comb connections:

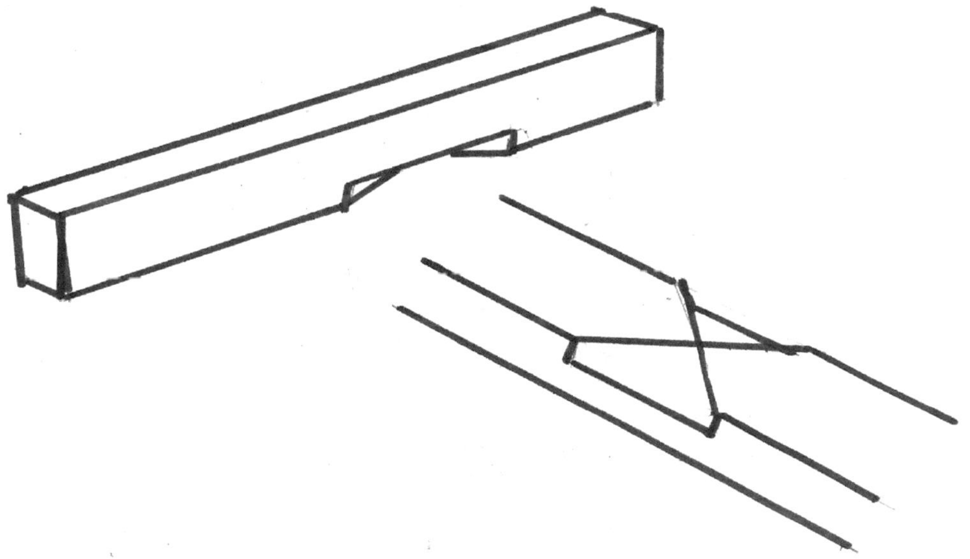

Kerve:

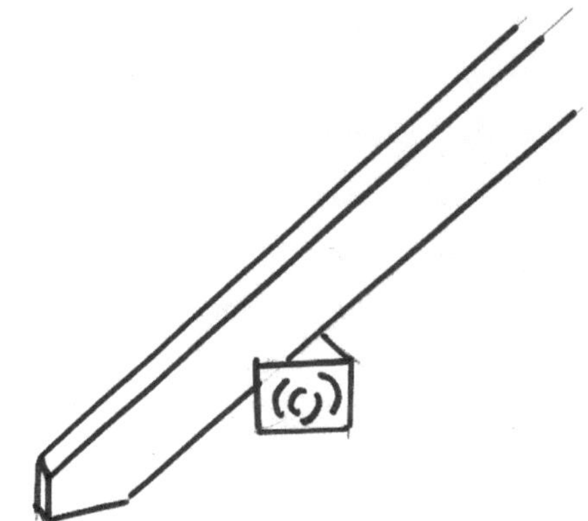

meshing:

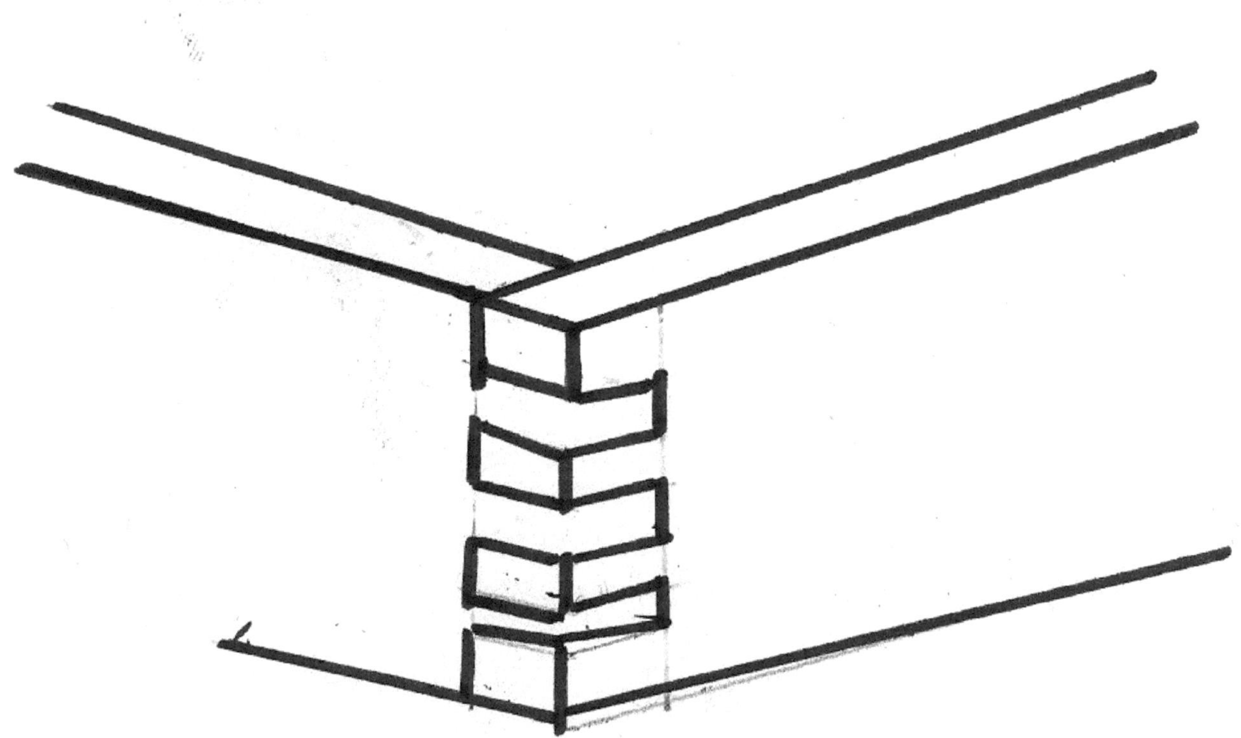

Row multiple connection: for such a connection, you can take a meshing or also take in a row at an edge at spaced-apart dovetail connections or dovetail... - Blade connection.
For example, for floor or roof extensions or as a connection of a terrace/porch with the main house roof and floor. It prevents (in this example) drifting by wind, sand compaction... of the terrace/porch from the main house. With the floor connection, the sheet needs to be oriented downward and with the roof the other way around, upward.

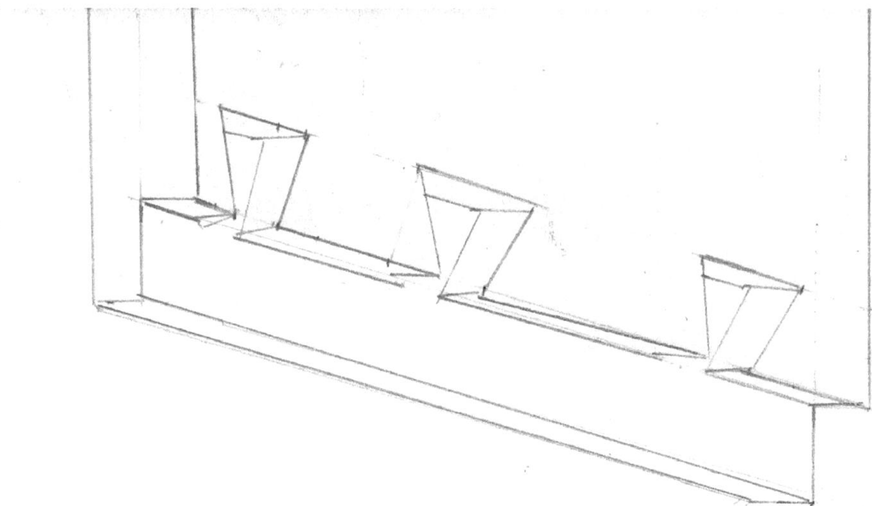

With every connection, it should be ensured that the wood is also able to withstand the pressure or pull. When connecting hardwood with softwood, the softwood is always the first weak point.

Staggered nail position for skirting nailing

If you nail or screw, you should always nail or screw against these forces when forces are applied. This prevents the nails and screws from leaning out. (Attach the soft wood to the harder one.)

Dovetail nailing and cross nailing

Offset row: these are nails that are nailed in a zigzag line, but not in a grain line. This is done so that the wood does not split along the nails. (for example, wallpaper or decorative strips)

Surface nailing: The nails are nailed in an offset line, regardless of whether they are beaten in the dovetail, cross or straight direction.

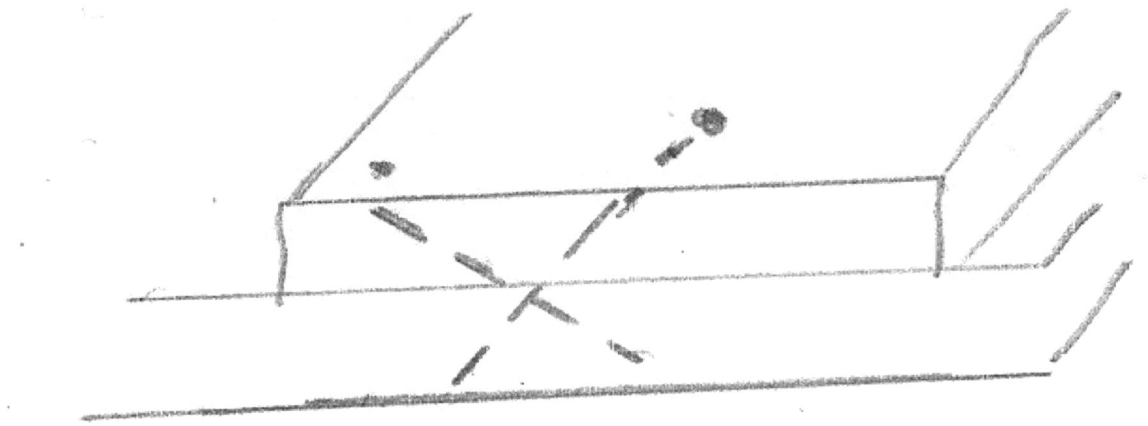

Cross nailing

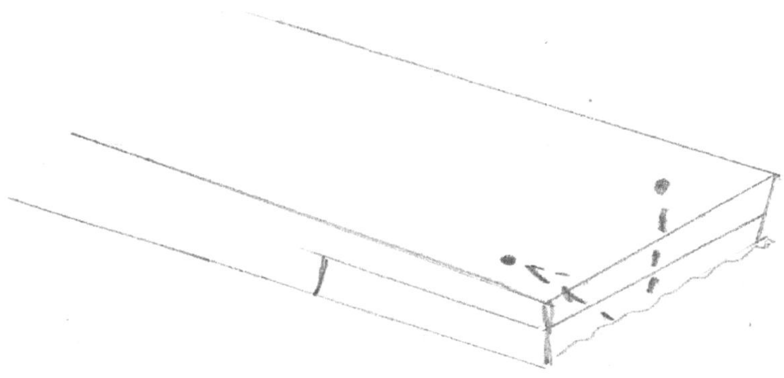

Dovetail nailing

Note: All exterior boards for animal housing should be smooth inside and on the longitudinal profile, not coated or impregnated with protective products, as well as have no protruding nails.

6.) A chicken coop for up to 6 chickens and a rooster

In advance: Of course, you can also build the stables, shown here, for more animals. The sizes correspond to the space requirements according to German law (as of January 2023). Of course, you can also build each barn larger to provide more space for the animals. Depending on the climate, however, the animals heat the interior with their own body heat. As more interior space also means more energy loss here. Because animals are just as happy about vitamin D (sunlight).

Do not heat the stables artificially in winter, if the chickens are allowed to live outdoors, otherwise they need, just as much "coats", as humans or some dogs.

Stables in construction law are referred to as mobile stables and non-mobile stables. This is not about whether you have wheels underneath, but the difference is whether they are transportable or firmly anchored in the ground. Transportable stables do not require a building permit (in Germany).

Up to 18 chickens can be kept in a residential area. A space requirement of 1m² is calculated per 3 animals. In my model, the water dispenser and the feed trough are in the outdoor enclosure. The chicken coop has a built-in floor. This prevents the fox from digging a tunnel into the stable as a night hunter. The nest and the sharpening stone should be in the stable, as the chicken wants to retreat for laying and / or brooding and the sharpening stone dissolves when not protected from rain. The stable here is 1x 2 meters tall. (Free run not included)

Roosters like to fly up somewhere to proclaim their empire from there. This can be from the stable roof or a post that is fixed in the run.

Sketch when done

Material wood stable:
- One plate 2 meters x 1 meter (bottom plate)
- 15 boards 2 meters length 10 cm width
- 20 boards 0.95 meters length 10 cm width
- 8 boards 0.8 meters length 10 cm width
- 1 board or plate 1 meter length 40 cm width + 10x 40cm roof slats
- 1x 40x40cm panel or (6x, 10x40 cm boards) as door
- 7 roof slats 1 meter long
- 6 roof slats 10 cm length

You also need:
- 7 Hinges
- 7 small angles (they serve to fix the walls on the base plate so that they do not slide back and forth)
- 2 padlock hinges and 2 bolts or mandrels (used as door lock)
- > 276 pcs. Nails 4 cm
- 42 wood screws 4 cm
- Chick wire mesh 6 meters long 15 cm wide (If you want chicks, otherwise rabbit wire mesh is enough, against egg thieves...)
  - Chick wire mesh 1 x 2 meters and
  - Chick wire mesh 2 meters x 1 meter from the 6m cut
- A roof plate: corrugated plastic (can best be cleaned from chicken droppings)
- Clamps or staple-shaped nails to clamp the wire mesh tightly

Cavity
2 boards 30 cm long and 10 cm wide
2 boards 25 cm long and 10 cm wide
1 plate 30cm²
20 small nails

hen roost
1.50 meter long round wood (wooden posts around 5 cm diameter)
6 roof slats of 20cm length + 2x 90cm
13 screws 6 cm

The board thickness is 2.5 cm for all boards

Tools: hammer, hand saw, screwdriver, staple monkey (from the tool shop not office store.)

Basic rule: Thin material is always nailed or screwed onto thick material. Turn over nail tips on the other side.

Roof slats please choose on average 4cm x 3cm

Step 1
Screw the base plate onto the 6 (10 cm) pieces of roof slats. Keep the pieces upright.
2 screws per roof slat piece. (These are the feet of the stable).

Step 2
Place 2 pieces of roof slats of one meter at a distance of 2 meters and nail the 2
meters long boards on them so that you have a 2 x 1 meter board wall. Place another
roof slat of one meter underneath in the middle and nail the boards to this slat. 3 Nails
per board and slat point are calculated. 1 Nail vertical and the two outer ones as
dovetail nailing.

Step 3
Nail 9 boards of 0.95 m also on a 1m roof slat and 9 more on another roof slat. Then
connect these two walls each at one end of the large wall. Leave 10 cm between the
top two boards and the rest. Here, the wire mesh is clamped against from the inside.
(see picture)

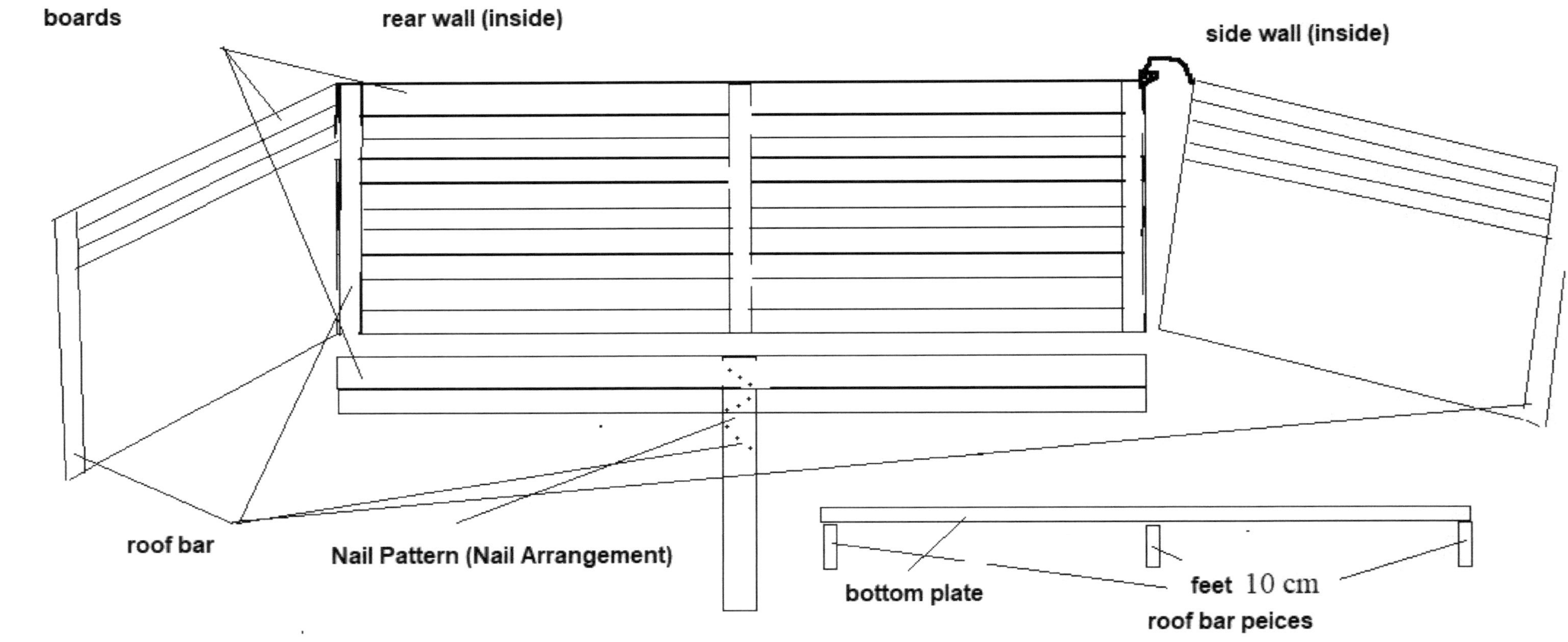

boards
rear wall (inside)
side wall (inside)
roof bar
Nail Pattern (Nail Arrangement)
bottom plate
feet 10 cm
roof bar peices

Now place the 3 assembled walls on the base plates and fix them with the angles. The lower ends of the roof slats with screws onto the base plate.

Front face:
The front consists of 3 sections. 2 wallsections and the door.

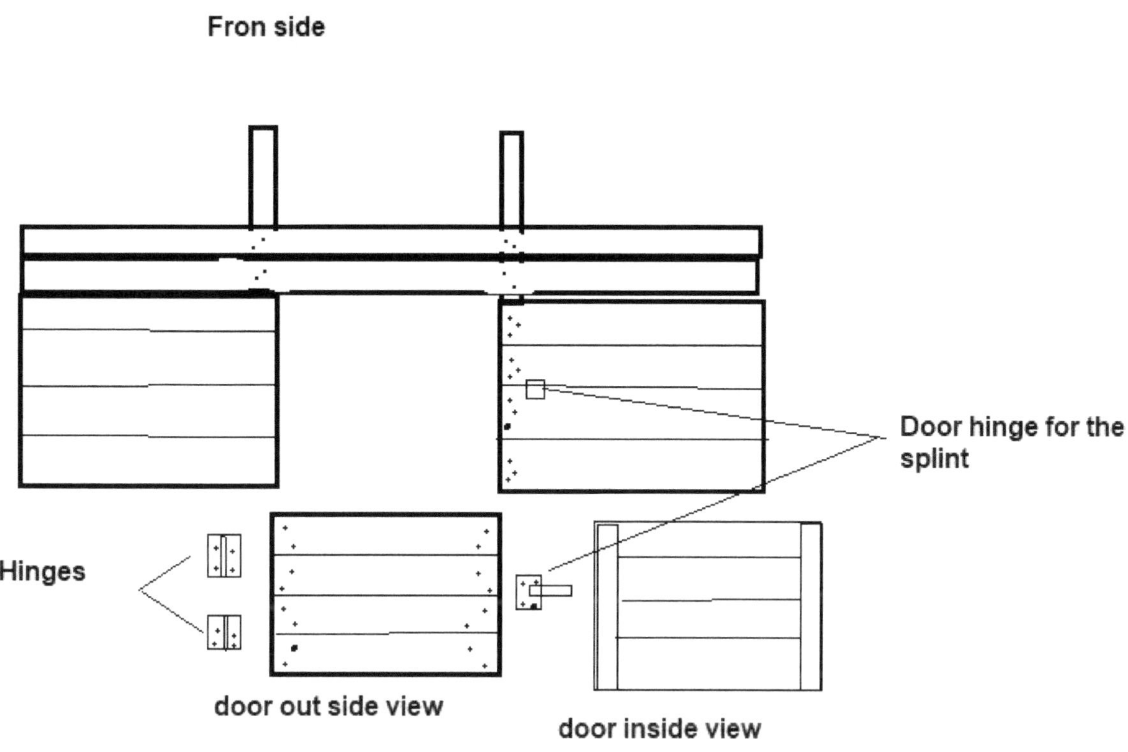

First, you nail together 4 of the 0.8m boards, as with the side walls, and connect them just as you connected the side walls to the back wall. Fix the new roof strip to the base plate with the angle. Repeat this on the other side. Via the 4 boards you now nail two 2 meter boards so that the 2 small front parts are connected and leave again 10cm free for the wire mesh strip. Nail 3 more 2 meters at the top to close the wall. Now clamp the wire mesh strips into the open strips all around. Now you should have an empty stable without door and roof and a 10 cm wire mesh windows.

For the door we now nail the 4 boards (40cm) onto the 2 other 40cm boards, so that a solid square of boards is created.
Now screw on 2 hinges on one side and the lock hinge on the other side. (See drawing) Now screw the hinges to one front board side so that the stable now has a door.
Now close the door so that you can fasten the lock hinge eyelet to fit exactly.

The nest

Nail the intended parts together into a "drawer" and place it in one of the front corners of the stable. (So you can reach it to get the eggs out of the nest.)
Hen roost
Screw the small roof slats together in a cross shape so that they have 5 cm on one corner and the rest on the other corner. At the small angle, you then insert the rod and fasten it with a screw (or a cord (optional)). One cross at each end and the third in the middle. As a stand stabilization, screw them against the two outer crosses at the bottom end of each one of the 90 cm roof strip, centered.
Now you can put the chicken roost in the barn at the opposite end, so that the roost is not above the nest. Chickens make their toilet where they are, right away. Therefore, the rod above the nest, the nest would not stay so clean.

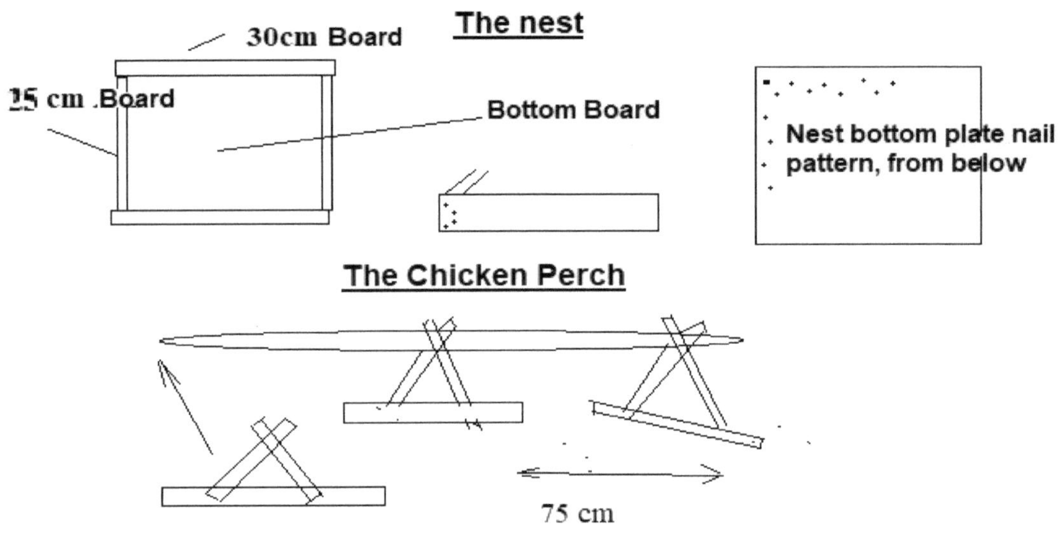

The chicken ladder:

For this purpose, the 40cm wide board and the 10, 40cm long roof slats are needed. Distribute the roof trim pieces on the board and nail them there.
Then connect the board with 2 of the hinges on the down side of the plate to the lower side of the floor plate in front of the stable door. (Due to the hinges, the unevenness in the spout does not matter, the angle of the chicken ladder optimally adapts to it through the hinges. This will also make it easier for you to once again level out the holes that a chicken might have dug.

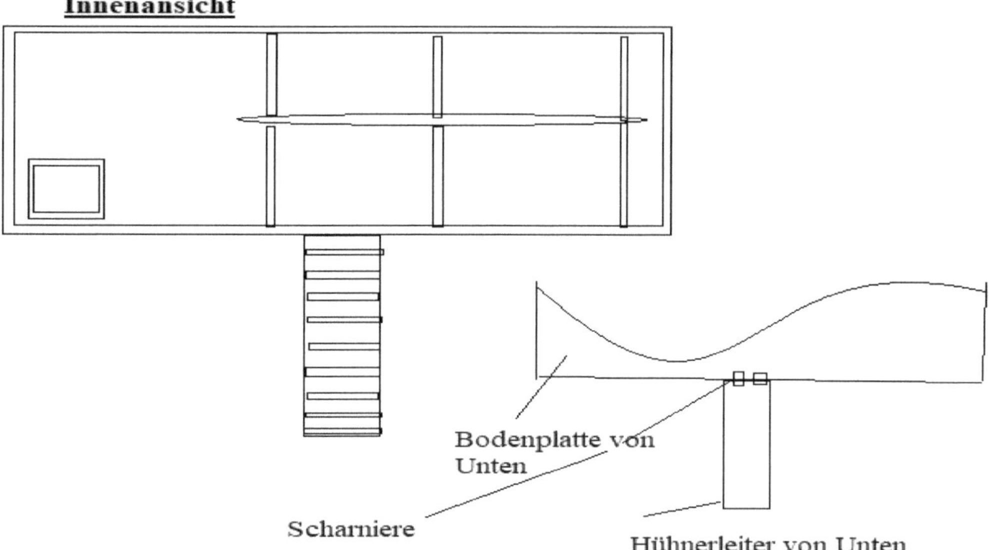

To the roof in advance the following thoughts.
The simplest method is, of course, to screw the roof plate onto the ledge points of the stable. BUT, I have planned it here so that you can open the roof. For example, for better cleaning, the nest box and the chicken rod can be removed. You can then sweep the barn with a broom from above after removing the straw or take out the eggs from above with a trowel if your arm is not long enough.
To do this, screw the 3 hinges on the rear longitudinal side of the panel and the door lock hinge on the other longitudinal side of the panel. Place the plate on the barn and THEN screw the 3 hinges to the back side wall of the barn. On the front side, place the lock eyelet to produce a perfect fit. This sequence reduces the risk of bursting during screwing the corrugated plastic plate. In addition, it is recommended to pre-drill the holes on the plastic plate. However, the drillhole should not be larger than the screw head. It is best to attach the hinges to the plastic plate with screws, nuts and washers.
For this method, you still need the screws and washers in addition to the listing.
(Personally, I chose this method at my stable.)
As with the door, you can now fit the lock eye exactly.

Now you have 1x 2 meter and 2x 0.95 meter boards left. These boards are intended to be hung on the outside in front of the wire mesh. Or you mount them with additional hinges to fold them down in front of the mesh which will prevent cold stress during the night, winter or storm.
However, there are also some that clip the simple bubble wrap in front of it in winter. In summer, however, you should remove the foil to avoid heat stress.

Since you can freely choose the window closing method, except the boards, the materials above are not listed

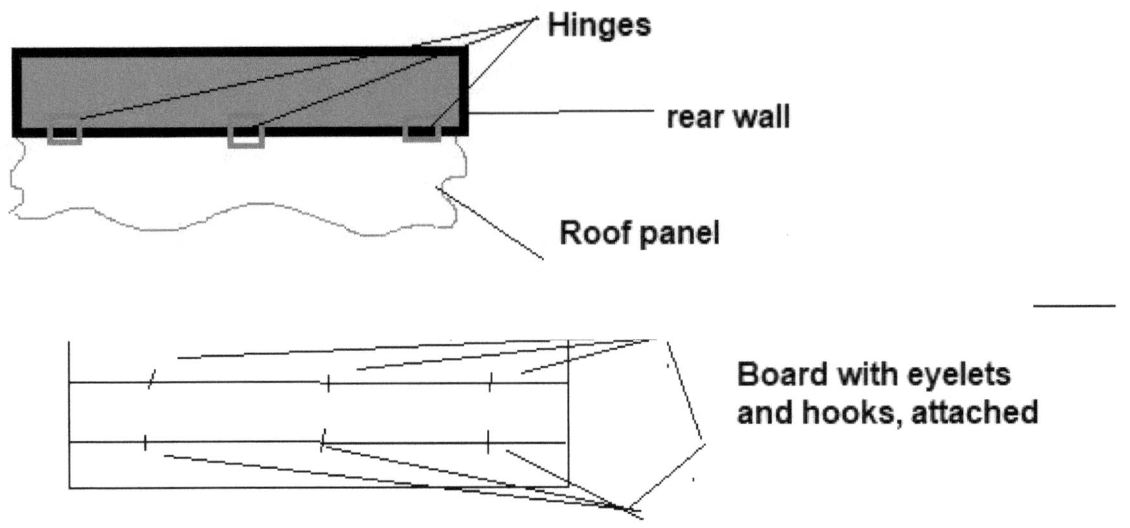

Hinge method:
7 hinges (3 for the front, 2 for the right and 2 for the left)
7 Eyelets and hooks.

Mounting method:
14 eyelets and 14 hooks.

Air cushioning method
4m x 10cm Foil and Clips

Note: If you choose the roof opening method, you will need 3 more stones to put on the roof so that it is not blown away in storm or strong wind.
An additional night protection against predators and hunters in the roof opening model, you should additionally use a net or wire mesh, hooked in nail hooks so you can roll it up when intending to open the roof.
The net/wire mesh field should then be slightly larger than 2x1m. The net can be permanently clamped to the back wall, a distance of about 10 cm is recommended. This means you need 20 clamps and 40 hook nails. Experience has shown that the net is not hung out every day. For complete interior cleaning or removing the chicken rod, however, this is the best method.
In the roof-closed version, you can also fasten the chicken rod to the ground – screw it on. The nest box can also be removed through the door. An ordinary fire hook, garden hoe or the like can be used to pull the nest box.

7 Rabbit house

Before you build a rabbit house, you should realize how many rabbits you want to buy and whether you want to breed them or not! Do you have the place for it?

Here is a space requirement table

However, this is with minimum free run space. In this example we build again 2x1 meter base in which you can hold 4-5 dwarf rabbits but only 2 large rabbits (see chapter 21, German animal space requirement law). In pack keeping (shared housing) (so no individual bays) it is recommended to keep only one pubertal or older male. Otherwise, they may kick and seriously injure themselves and can even seriously injure each other, especially in the mating season.
(I got this information from a small animal breeding club.)

Rabbit hutch front view

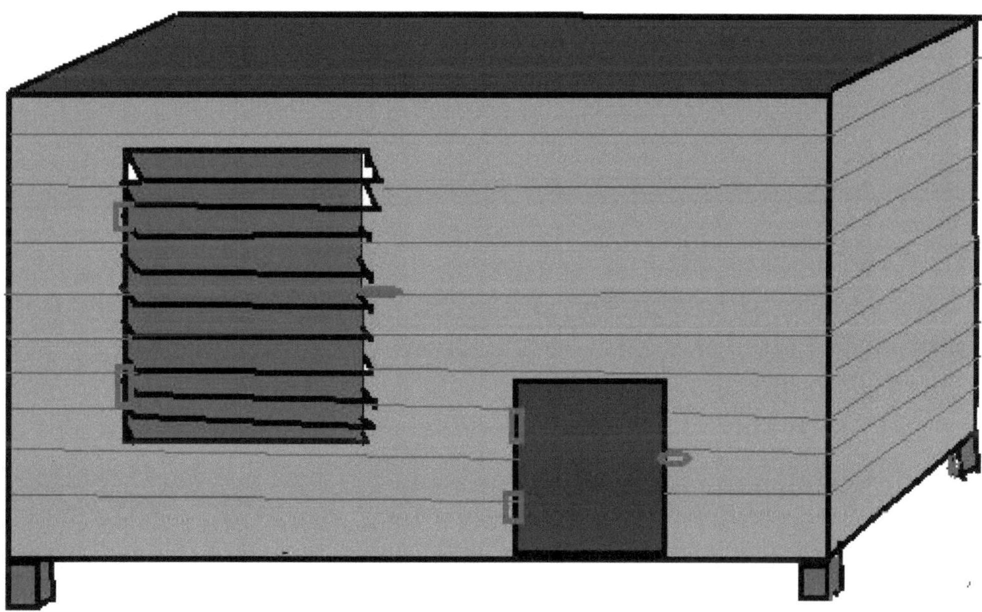

The total space requirement with spout for these numbers of animals is at least 6m².

Here the door is locked against night hunters. If you have a spout that protects against them, you can perhaps also hang a felt leather in front of it to keep the stable heat.

Number of Rabbits	minimum space requirement
2	6m² floor space (e.g. 2x3m)
3	7,2m² (e.g. 3,60x2m)
4	8,4m² (e.g. 4,20x2m)
5	9,6m² (e.g. 4,80x2m)
6	10,8² (e.g. 5,40mx2m)
7	12m² (e.g. 2x6m or 3x4m)

Materials Requirements

9 x 2m boards 10 cm wide
16 x 0.95 m boards 10 cm wide
13 x 60 cm long, 10 cm wide boards
9 x 30 cm long, 10 cm wide boards
4 x 1.1m long, -"-
1 x 1.4 m long, -"-
4x 70 cm long, -"-
3x 40 cm long, -"-
6 x 80 cm long roof slats 3x4 cm
and 4-6x 10 cm roof slats as legs (if they are thick, it may be better to take 6 legs).
24 washers approx. 3-4 mm thick 24 wooden screws 25mm the head of the screws
must not be able to slip through the washers.
6 hinges and 2 door lock hinges including wood screws 2 spikes
6 small metal angles and 24x 30 cm wood screws (12 optional with 48 wood screws
of this length for the legs with hinges)
16 – 24 wood screws 3.5 cm for the legs and roof panel (if you do not take angles for
the legs) otherwise 8 – 12 wood screws 3.5 cm
1x 62 cm² rabbit wire
4 Aluminium flat brackets (optional for the window drawer and corresponding
number of wood screws).
2 panels 2 x 1 meters as floor and roof panel.
1x 30 cm² door panel or (optional leather felt)
> 332 Nails 3.5 – 4 cm
Clamps for clamping the wire mesh
Board/plate approx. 50/80 cm x 30 cm for the entrance ramp for hopping up/in of the
rabbits.

Tools: hammer, cordless screwdriver, craft clamp witch, tape measure, pencil or tear
nail, hand saw and planer or chisel and wooden hammer/rubber hammer. (Brush and
weather protection, suitable for animals), spirit level, metal file

rear left corner view with floor panel inside

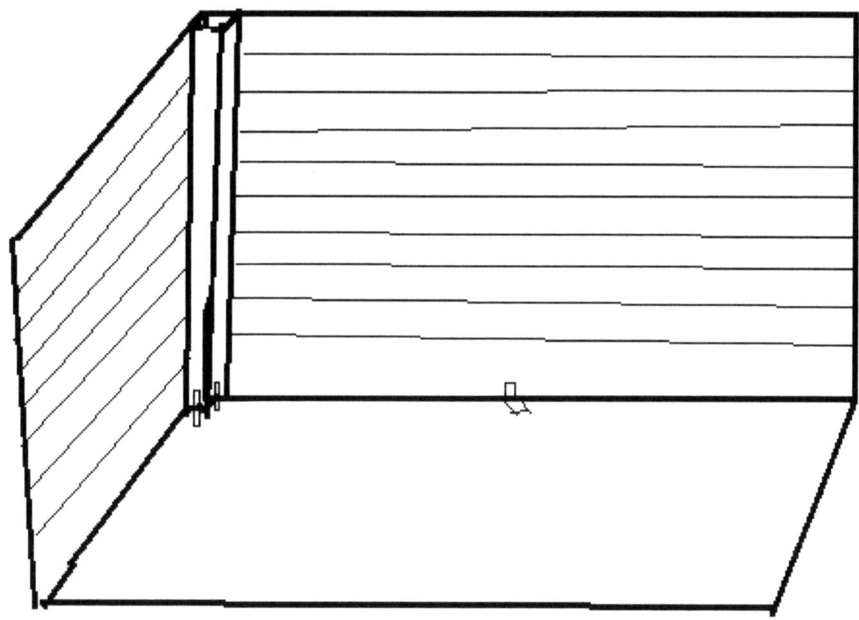

Step 1

Screw the base plate onto the legs. That is, the screws from the plate through the plate into the legs (10 cm long pieces of roof slats).

Or option 2: Attach the legs to the bottom of the baseplate using the angle and wood screws. If the tips of the screws on the top of the base plate look out, you file them down with the metal file. So that there is no risk of injury for you or the rabbit.

Step 2

Place 2 of the roof slats at a distance of 2 meters on the floor and nail 8x 2m boards on the roof slats together to a wall, 3 nails per board end in zigzag line.

Then place a third roof slat exactly in the middle under the boards and nail them as well.

Then screw this wall onto the base plate with 3 angles. One angle per roof slat.

Place another roof slat on the floor and place the 95 cm long boards (8 pcs.) with one end on the roof slat and nail them together again to form a wall. Repeat this with the other 95cm boards. Now connect these as right and left wall with the rear wall attached to the base plate. So that the roof slats are pointing to the open stable side, inside.

Nail the free side of the roof slats onto the roof slat of the rear wall and screw the new roof slats onto the base plate with an angle.

Step 3: Front side Step 1

- Nail the 9th 2 meter board to the top of the roof slats of the left and right sides.

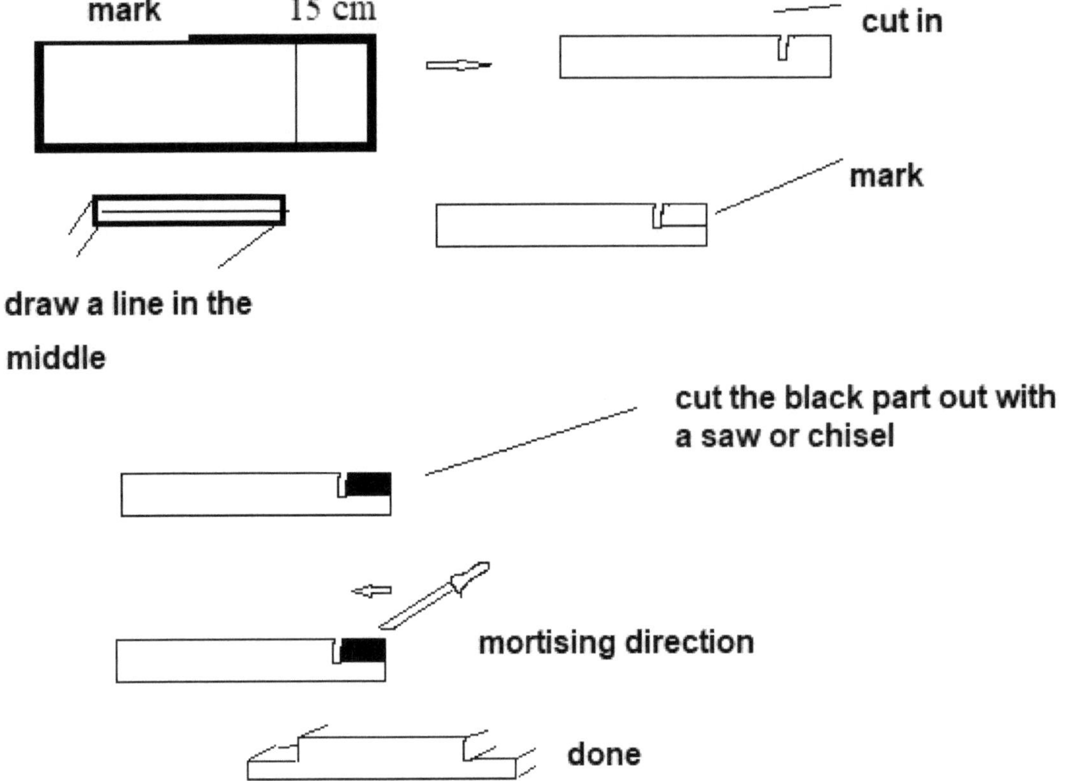

- Now take the remaining roof slat and place it in the middle in the front. Nail the board to the roof slat and fasten the roof slat to the base plate using angle and screws.
- Now nail the 4 - 1,1m long boards on the roof slats on the right side. (To the left will then come the window.)
- Nail the 1.4m long board onto the roof slats at the bottom on the left side.

Front side Step 2

Build the window frame according to the drawing and your choice.
When selecting the square blade joint, see Wood joints, page 13 (Blade butt joint).
Place the two 70 cm long boards on a workbench/table and draw from the ends, 15 cm long a line of width side. So far, the bumper sheet must go into the board. Cut the board half deep along this line. Now clamp the board upright in a vice, divide the board cross section in width with a pencil line into 2 equal parts and connect the line with the ends of the incision depth. Now use chisels and wood/rubber mallet to punch out the marked area. Or you saw this part out with a fine saw.

Now do the same with 2 planks 60cm long, saw or cut out here **only** 10 cm.
Assemble this into a frame (nail). Turn over nail tips on the other side. With the long boards you have an overhang of 5cm on both sides. (This is for nailing it to the wall on the inside.)
Now take 6 of the 30 cm long boards and 1 x 60 cm long board and nail the 30 cm long boards on the 60 cm long board to a wall shape, going 5 cm from the board ends inwards. Then, after drawing, nail the window and the wall piece together.

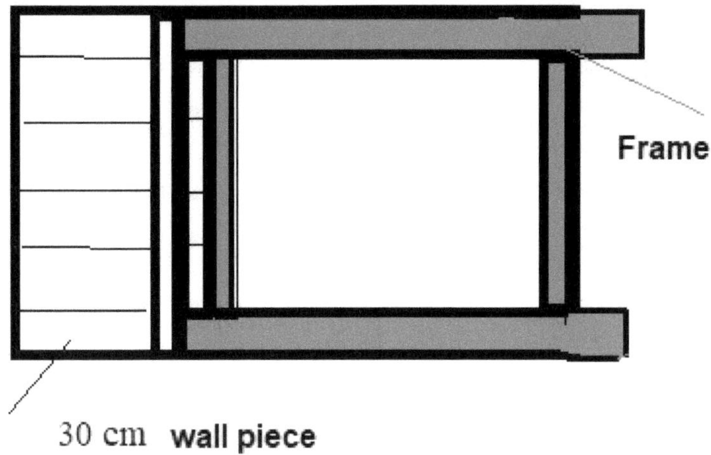

Frame and 30cm wall piece (interior view)

Now clip the rabbit wire mesh panel into the frame.
Now nail the wall piece part onto the roof strip. And then the frame over space on the other side together with the 1.1 m wide board. (Attention here, it can bounce a bit when nailing. Lighter hammer blows may take more time, but then they more are helpful.)

Option 2
Place the 70 cm long boards and 2 boards 40 cm together to the frame as shown in the picture and screw the bracklets or angle plates to join the boardangles.

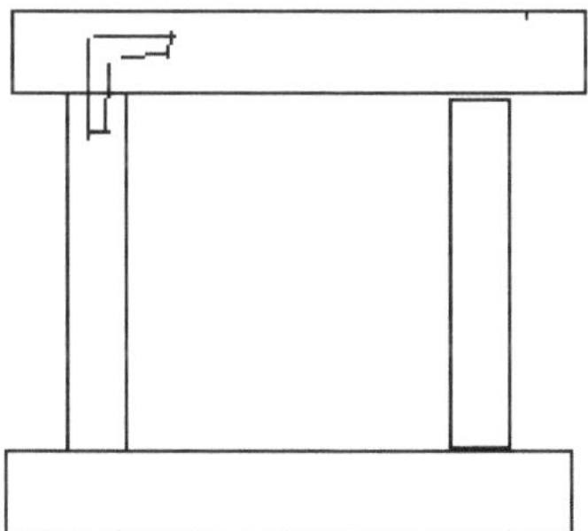

Now proceed as with the blade butt joint.

The front wall should look like this until now.

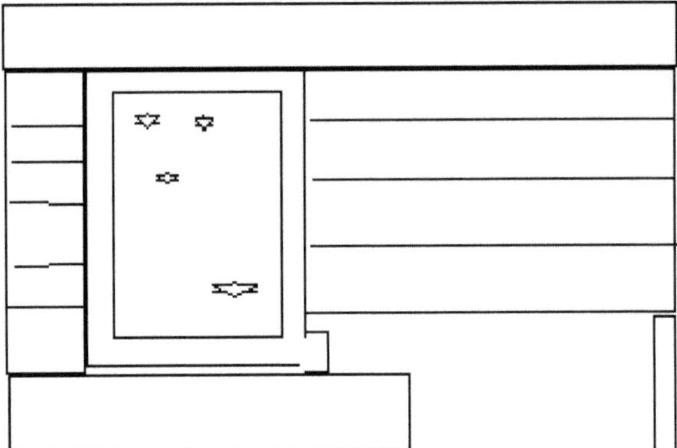

- Now nail one of the 40 cm long boards on the inside, upright and flush with the 1.4 m board
- Next take the last 3 boards 30 cm and 1 board 40 cm and nail it to a wall piece. Allow the 40cm board to protrude 1 cm. (as a door stop) and connect this piece of wall again to the roof slat.
- Now, apart from where the door enters, they have a hole of 60 cm in length (window frame connection and door frame board.) which can be closed with one of the 60 cm boards.

inside view

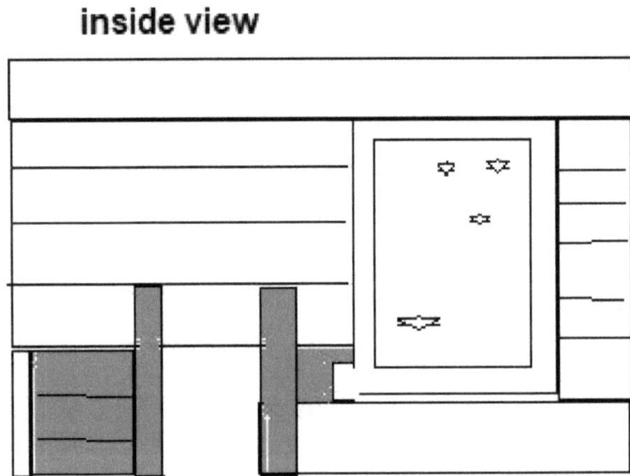

Screw the two hinges onto the door panel and the door lock hinge on the opposite side and insert the door into the front wall. Adjust the thorn eye fitting withe hinge to the wall and then screw it to the wall.

The window shutter

Place 2 boards 60 cm right and left as frame side and nail them on both ends to 2 boards 70 cm, which creates a frame. Nail a cross pattern.
Now screw one screw on the right and one on the left with 2 washers each on the lower side of the frame.
Then screw the hinges to the frame as shown in the picture.
Now you distribute 6 x 60 cm boards evenly into the frame. The lower side of the board should rest on the screws of the underlying board, so that the screws and washers prevent the boards from tightly resting on the lower board, thus creating an air slot.

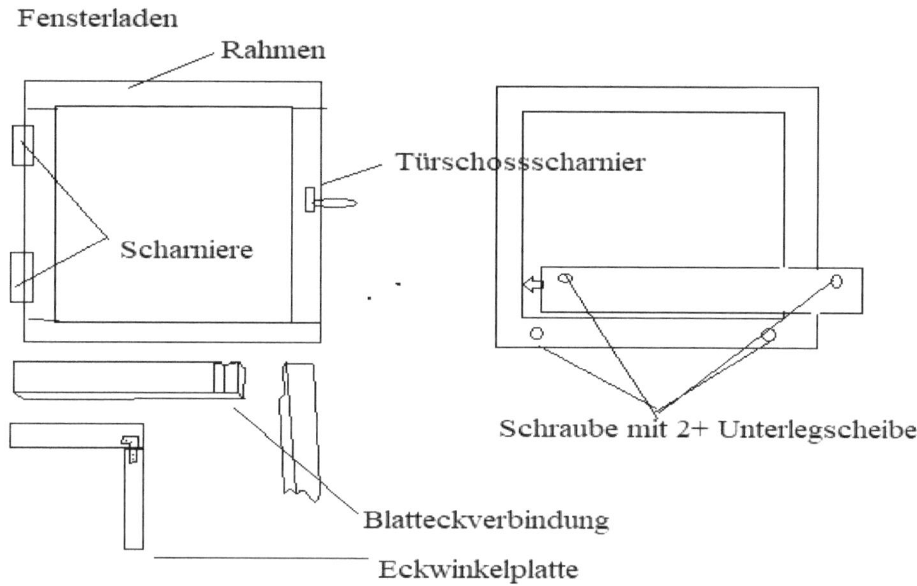

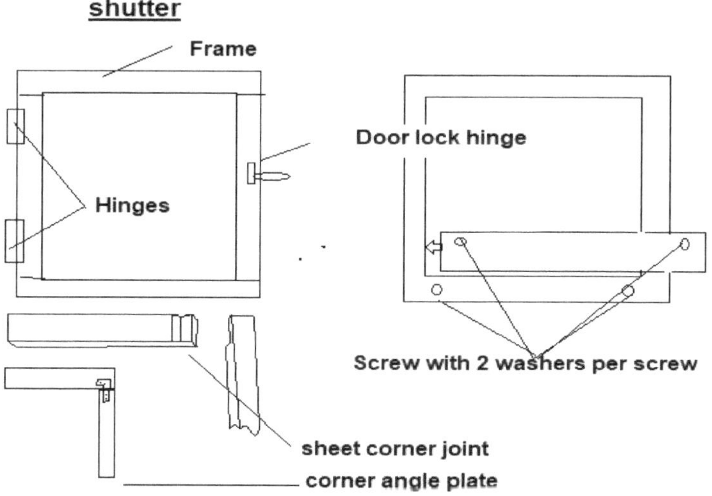

8. **<u>Dog house made to the dog measure of the dog</u>**

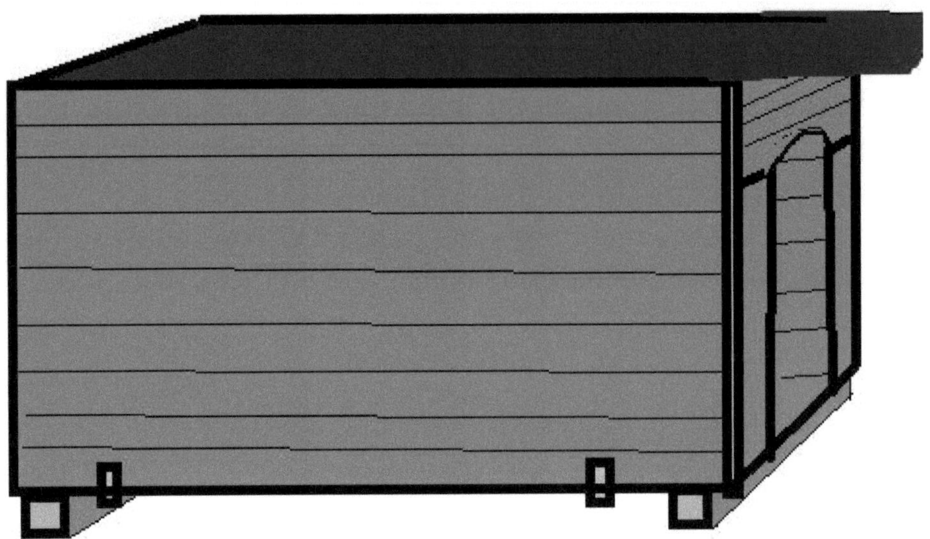

Your prestige does not matter to your dog, they do not need a large villa and they are not on a Zen Tripp. They want to retreat and sleep in it, protected from the cold and rain. Therefore, the roof should be closed (waterproof). The house should not be too big or too small.

The height of the house should be the shoulder height of the dog times 1.2 to max. 1.5, whereas the entrance only 0.8 times of the shoulder height.

Calculate the length of the dog times 1.5 for the length of the construction.

The width is equal to the withers of the dog but not wider than the length of the building.

Why is the roof flat on my plan?

I have seen the dogs of my parents loved to climb/jump on the roof and lie up there in days of good weather. But perhaps it is also because a place above the immediate home is so much easier to guard.

A dog is an animal that has taken over the role of guarding the area of its pack for food and not a lap dog. Some dogs take on this role (as well), but not every dog.

Don't let someone make you feel bad when your dog sleeps in his cottage in the garden. It is, literally, his room.

They are still worthy of being cuddled and definitely deserve a part of the time of the pack (family).

Placement is also very important.

If possible, prevent the hut from being blown by all 4 winds. Build a windshield on one side or place the hut beside a wall.

If you optionally build a small canopy, the dog has a shade in the summer against the accumulated heat in his (room) hut.

Step 1
Measure your dog and calculate the size of the future hut.

In this example I take a shepherd dog; shoulder height 60 cm, length 1 m, withers 60 cm.
This determines the size of the hut.
Height: 72 – 90 cm
Length: 1.2 – 1.5 m.
Entrance height 48 – 50 cm
Width: 60 cm – 1 m

The floor of an outdoor animal housing should never be directly placed on the floor, to prevent rainwater from running into the housing.
A full-grown dog can weigh quite a bit. That's why I use a squared piece of wood as legs.

Material requirement:
- 18x 1.2m to 1.5m long boards
- 9x 0.65 – 1 m long boards
- 2 squared woods 60cm – 1 m
- 1 wooden board 60 cm – 1 meter wide by 1.2 – 1.5 m long

Optional 7 – 10 x boards 1.2 cm to 1.5 m long boards for the floor
- one roof panel, wood 70 cm x 1.8 – 2.1 m
- 2x 90 cm + 2x 1 m wide boards or 2x 90 cm long boards + one plate 45 cm by 1 m (entrance, sheet processing)
- 4 boards in length = height of the hut minus 5 cm 4 metal brackets + screws
- >254 nails 6 – 6.5 cm long

Tools: hammer, hand saw, wooden file, screwdriver (cordless screwdriver), metal file, sandpaper, jigsaw

If you have chosen the floor from boards, nail the 7 boards of 1.2 – 1.5 m as an area on the squared timbers. There are 3 nails per board end to squared wood calculated. Nail a zig-zag pattern. Allow the squared timbers to stand about 5 cm inwards. (Not flush with the hut entrance or end.)

Step 2
Nail 9 boards of 1.2 to 1.5 m length each, on 2 of the height boards on to a wall and repeat this with the other 9 boards of 1.2 to 1.5 m length.
Flip over the walls and flip over the protruding nail tips. So that they are folded into the wood.
Attach the two walls to the base plate using angles and screws.

The angles are fixed on the outside of the hut so that your dog does not scratch himself on them. Grind down any protruding screw tips with a metal file.

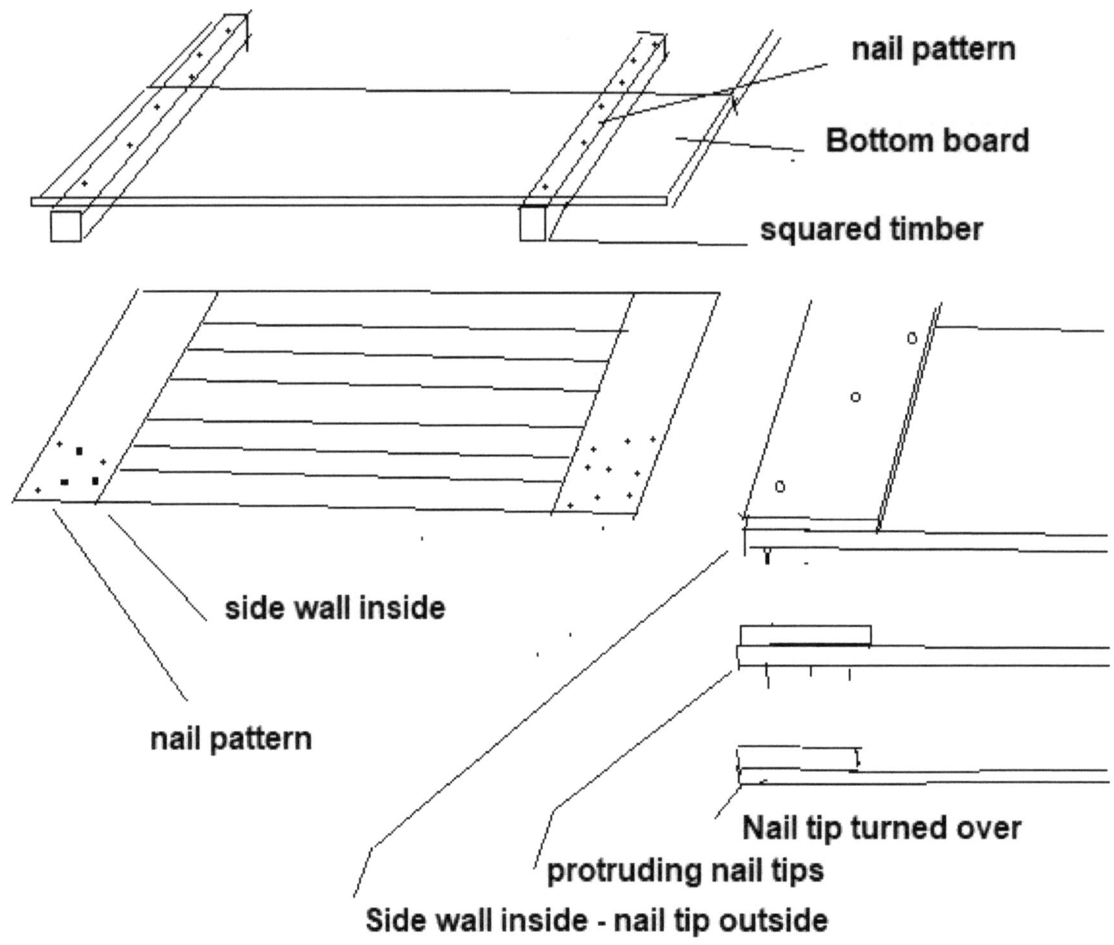

Step 3
Now nail the boards for the back wall 9 x 1m at the back of the hut. It is advisable to start at the top. This stabilizes the side walls right at the beginning.
Make sure that no nails protrude into the interior.

Step 4
Option 1 Nail the 3 meter board and the 2x 0.9m long boards together into a flat U-profile for the entrance wall.
On the fourth board mark 10 cm inward of each end and about 5-7 cm from the top, as shown in the picture below and connect these 3 markings to form an half circle. Saw this sheet and file or sand it smoothly. It is best to saw the area to be sawn out several times with the hand saw and then finely cut out with the jigsaw.
Then you nail this board with the bow pointing down onto the entrance wall. Turn the wall over and flip over the nail tips.

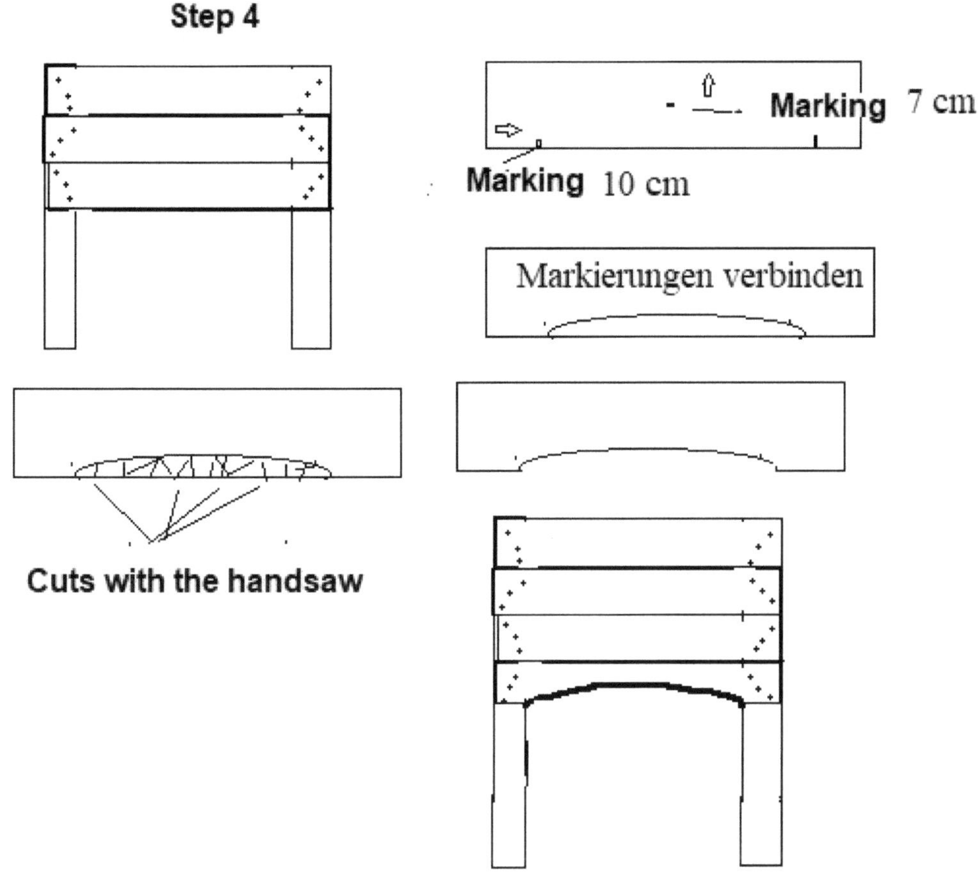

Step 5
Nail the finished front wall on the front and finally the roof. Again, make sure that no nail tips protrude into the interior.

If a nail has become crooked and the tip protrudes inwards or outwards, they must be pulled out again and another nail must be applied, slightly next to the same nail site, so that the new nail does not bend back into the old nail line and look out again.

9. **<u>Cat tree</u>**

Cat tree model: hanging mouse

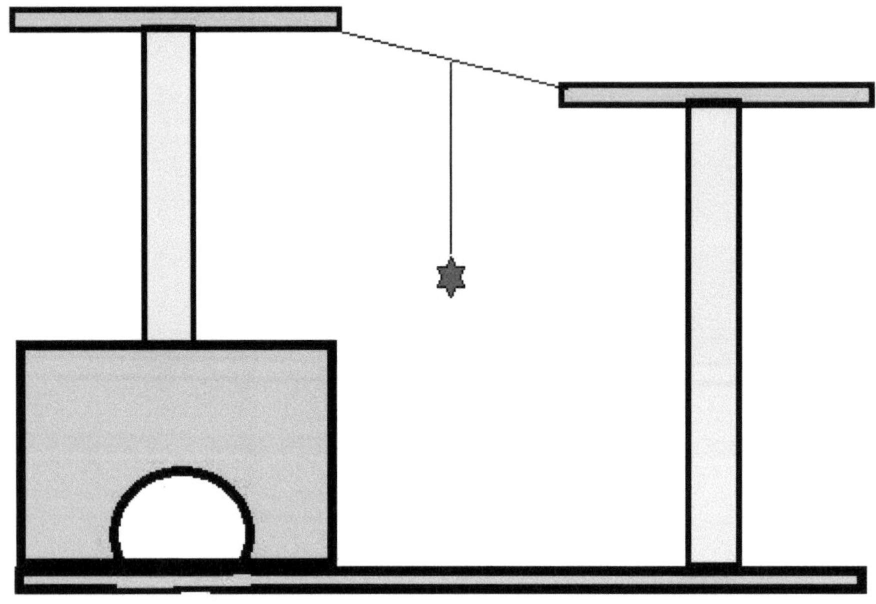

To be honest, I've never seen a cat tree where a cat tree is provided with a place big enough to sleep, where the cat can stretch out and sleep properly. Nevertheless, the cats love to retreat for a nap, climbing around and playing.
My cat has a large cushion on the bench (he loves my closeness) and individual components of a cat tree.
One question in advance: do you have a domestic cat or a free-running cat, or just a constant visitor?
Since my cat is a constant visitor, I also have a cat tent with fur, where she can "snuggle up" under my gazebo when I'm not there.
A garden cat, free-running cat will use a cat tree less often than a room or apartment cat.
Cats are explorers, athletic acrobats who are very active.

The outer surfaces can be covered with fabric or fur, to your cat this makes no difference, whereas fur within the cottage is more popular.
I choose 1m x 0.5m as the base plate and thus space requirement.

Materials Requirements
- One plate 1m x 0.5m
- 2 plates 35 cm²
- 1 round wood 20 cm long, diameter 5 cm
- 1 round wood 50 cm long, diameter 5 cm
- 7x 40 cm² fabric if you choose fabric outside and 6x 40 cm² fur or
13 x 40cm² fur
- 1x 1m x 0.5m fabric or fur for the base plate
- a string (package string or similar) 40 cm and a 5 -10 cm long
- Natural rope in thickness 1 cm, 1x 6.8m, 1x 15.75m (for scratching posts)
- a plush, plastic or rubber mouse approx. 5 cm in size
- >24 nails 5- 6 cm
- >6 groin nails with head
- 4 wood screws 6 cm
- Brackets for the bracket monkey from the hardware store

Tools: hammer, cordless screwdriver, hand saw, jigsaw, sandpaper and wooden file
and a stuffing needle

Preparation: For a 35 cm² slab section, saw out an bow like the entrance bow of the
dog house, mark at 5 cm.

When you have chosen fabric, you clamp the fur on one side on 4 of the 35 cm²
panels and the entrance and clamp the skins to the panels/board side. Now cover the
other side with fabric and the other boards/plate parts also on one side with fabric.
If you have chosen fur, then stretch out the fur there instead of the fabric.
Wrapping the round woods in two layers with the respective natural rope, then fix the
beginning and the end of the string with a groin nail. Before nailing the end, make a
knot in the end so that the rope does not dissolve.
Wrap the rope very tightly together. The round woods are the scratching posts of the
cat tree.

Step 1

Screw the roof of the cat house to one side of the 20 cm log
the outside of the roof to the round wood. (In the case of fabric, the fabric is on the
roundwood side) and a 35 cm² part covered on one side on the other roundwood end
with the uncovered side to the roundwood.

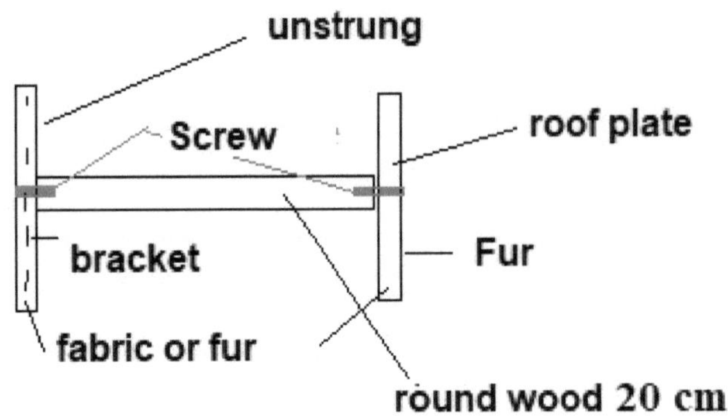

Step 2

Nail the 3 walls and the entrance together with 3 nails per adjoining edge. The fur
side inwards.
When you have covered the bottom plate with fabric, fix the last fur part on the
bottom plate left-centre with a few brackets.
Place the house walls on top and the bottom plate upside down. The small house
walls hold the inner floor skin firmly to the floor plate. To avoid wobbling, you place
an object under the other end of the base plate, with 35 cm height. This can also be a
cooking pot, for example, without any problems.
Select the nail area.
Two methods: you first put the house walls nailed together on the underside of the
base plates and mark around the walls inside and outside with a pencil (this is the
easiest method) or you enclose the base plate with a string on the outer walls of the
house and mark along the string.
Then nail the bottom plate onto the housing part. The nails should **not** look out of the
walls on the other side.

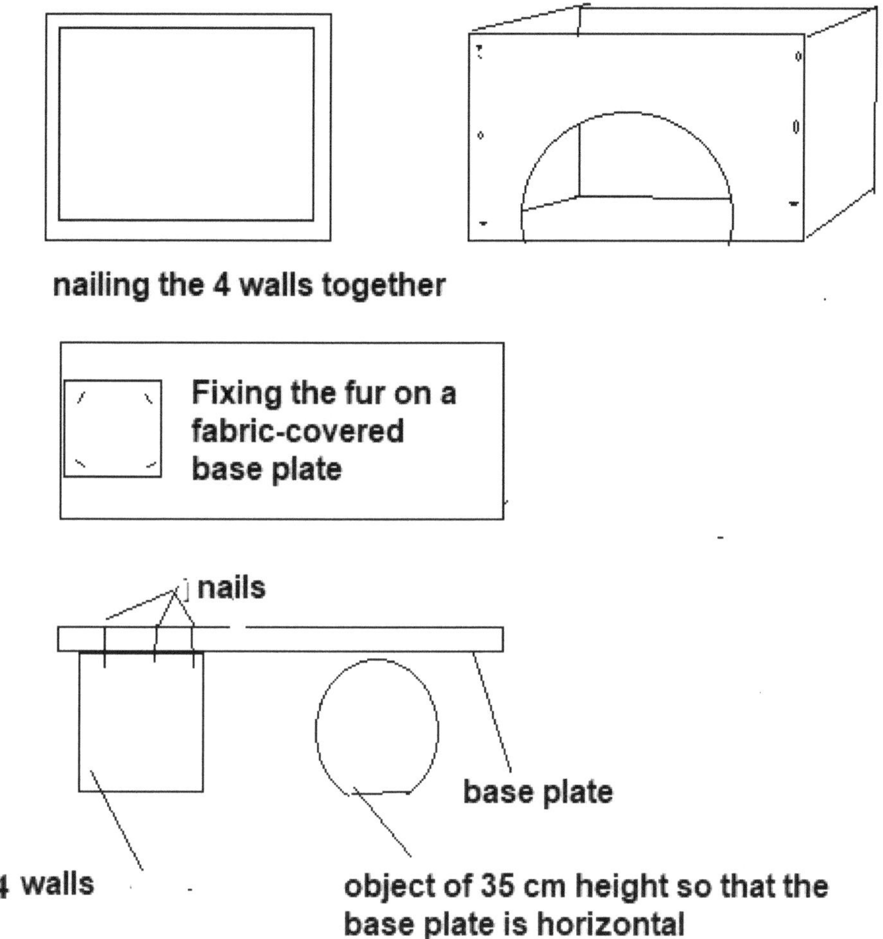

Step 3

Screw the remaining 35 cm² plate to the 50 cm round wood end, with the fabric or fur side facing outwards.
From the right bottom panel end, mark a point on the bottom in the middle 17.5 cm.
Then place the finished cat tree part on the side so that you can screw this scratching post part with the jumping platform to the bottom plate when marking.

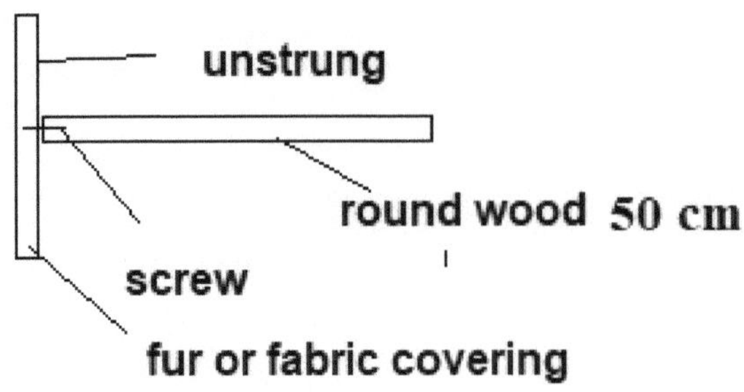

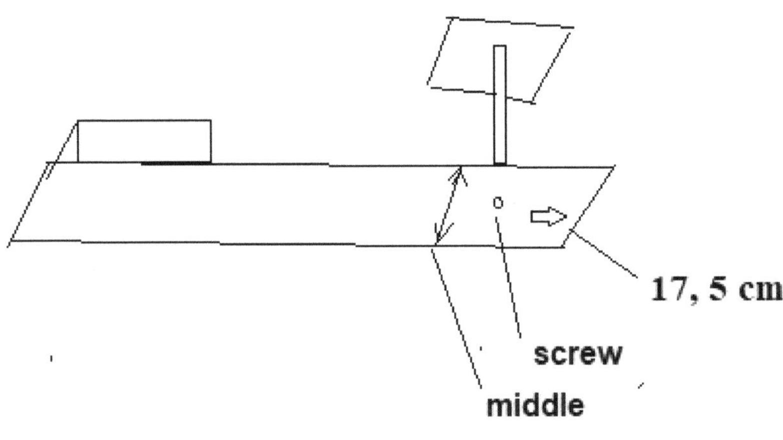

Finishing:

Place the cat tree so that the attached scratching post stands upright.
Now take the roof with the little scratching post and nail the roof on the walls. Make sure that no nails protrude to any sides.

Tighten the 40 cm string between the two spring plates after both ends were knotted to avoid dissolving and nail the ends with a groin nail.
Then you knot one end of the 10 cm long lace in the middle of the stretched string. With the stuffing needle, you pull the hanging part of the string through the mouse until you can knot the mouse.

If your cat then looks at the whole thing curiously, then blow against the string so that the mouse starts moving and leave the manege to the cat.

10. Bird cage for 1-2 birds

Wicker Birdcage

This is not supposed to be a simple box or something like that made of strips and wire mesh, which would also work. But you don't need an instruction book for that.

To make this bird cage, you need water in a tub. The willow rods must be wet-moist so as not to break during processing.

Materials Requirements
- Wooden disc 70 cm diameter (circumference 219.91 cm)
- thin wicker rods > 300 à 75 cm
- 3 round woods for the bird poles approx. 35 cm long and optionally 1 round wood for a swing
- a roll of flower wire
- a plywood board 5cm²
- Parcel cord
- 1 button or little hook
- Thread or sewing thread
- 2 small key rings

Tools: tub water (>80 cm), drill with wood drill in the thickness of the wicker, fine hand saw or blade saw, craft needle (old sewing or stuffing needle, can be bent), a small nail, pencil and hammer

Step 1
Soak the wicker until it bends well, but do not allow it to swell for too long.

Step 2

Slightly beat the nail in the middle of the wooden disc and knot the thread or sewing thread on the nail without cutting it off. Put the pencil through the sewing thread roll

or twine star and then unwind the star or roll so far that the pencil tip is 32 cm away from the nail.
Now draw a circle on the wooden disc with the pencil. Remove the nail, pencil and thread.
Natural willow rods are of varying strength. Measure the thickness of a willow rod and add 1 cm. Now make markings on the circle at a distance of this Length and drill a hole through the disc there so that you can put the thick end of the rod through. However, the hole should not be too large, otherwise it endangers the stability.
Put the thick end of a rod through each hole.
Then pierce the rod with the needle at half a centimeter and there push the flower wire through, as if to a continuous ring and knot the ends. Cut off the rest of the reel and pull the rods back so that the wire rests against the disc.

Step 3

Now braid 5 rods into the upright rods to form a basket wall and press them against the disc. You can knot the two rod ends with the flower wire for durability, but make sure that the knot ends are outside the cage.
Then pierce with the needle through the upper ends of the rods at the same distance, for example 1 cm.
Then pull the thread through and pull the tips together tightly. And knot the thread tightly together.
Walk the flower wire through one of the top extinguishers and tie the wire to the route. Braid the wire into 3 rings and weave the end.

Weave in 3 double rings of willow rods at equal distances from the disc and the tip and secure them with the flower wire.

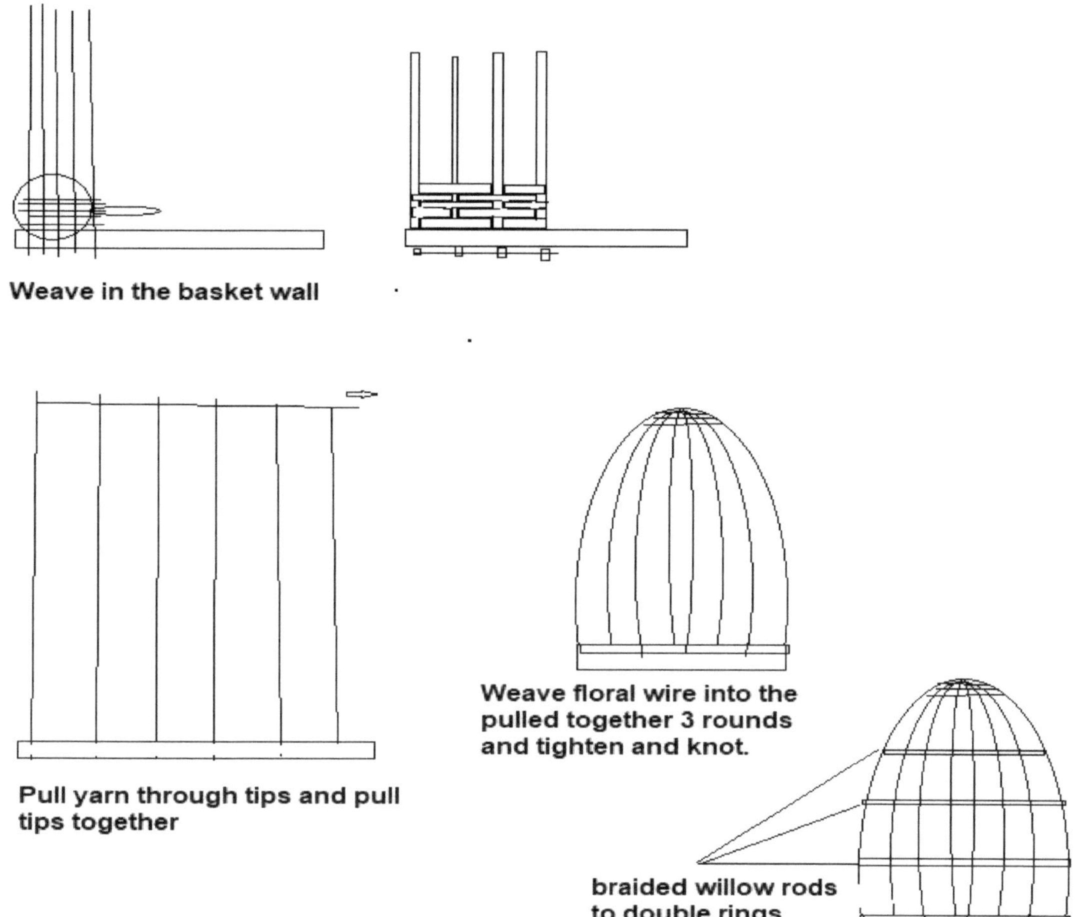

Step 4

Now saw, into the ends of the bird poles, so that you can stick a willow rod into the saw notch. With a fine saw or a jigsaw insert the rods into the rod struts so that they rest on one of the double rings.
The weight of the bird can thus not push the rod downwards and thus not push the rod out of the struts.
Distribute the poles at different heights according to your liking. Adjust the length of the bar.
Take the remaining pole that you have planned for a swing (optional). Drill at about 1 cm from the end, at both ends of the rod through the rod and pull through a package string about 20 cm long and make a knot in one end of the string and knot the other two ends over the cage from the outside.

Step 5

Cut a hole of 15 cm² in the cage wall with a hedge trimmer or strong kitchen scissors. Take 2 of the cut rod parts or cut them from the remaining, 2 rod pieces to make the upper and lower frame struts. By sewing a piece at the top of the strut ends with flower wire and a piece at the bottom strut ends.

Drill two holes in the middle on one side of the board and one on the opposite side. The two holes must be so large that you can screw in the key rings and rotate them well in the board. Through the other end you pull some flower wire or package string for a loop/eyelet.
Now sew the key rings to the cage so that they cannot slip on the cage and the board, when folded to shut, closes the opening. On the other side, you sew the button (you can also take a hook) to the cage with the flower wire. So that you can close the cage door with the loop over the button or hook.

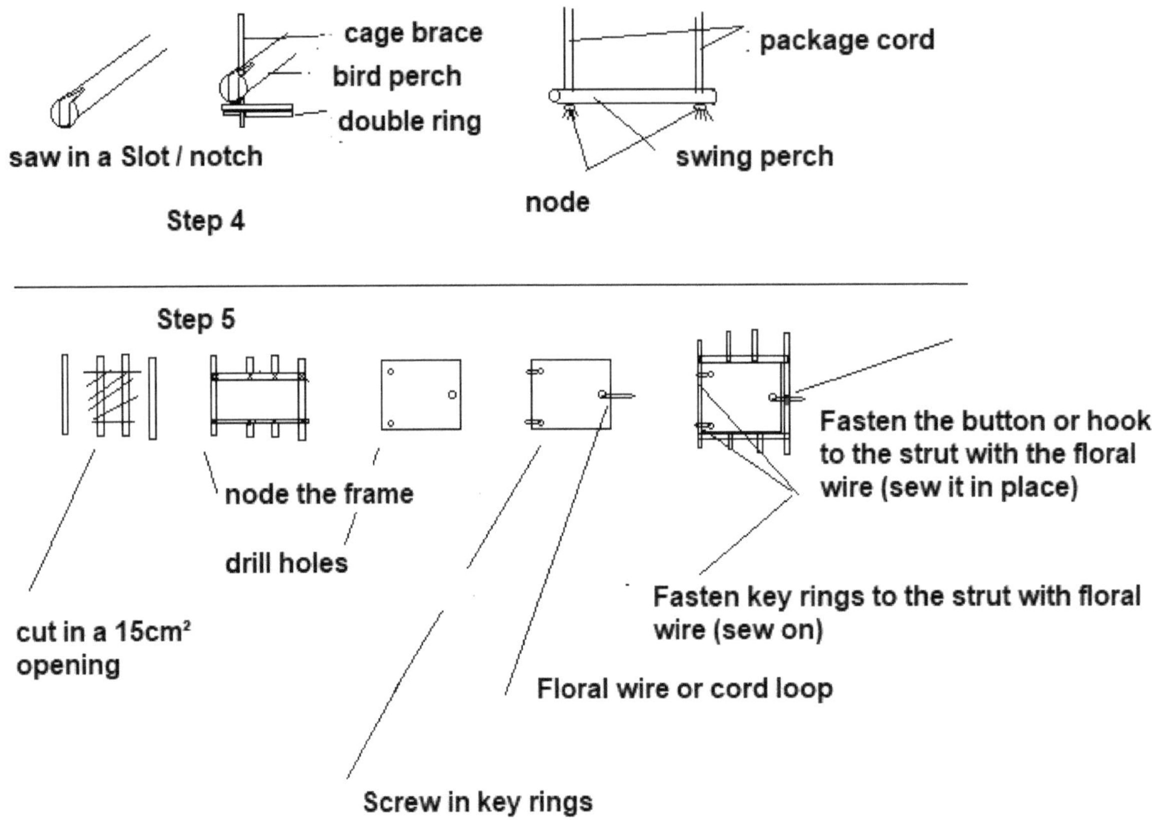

You need for the birds still a water dispenser and a food dispenser as well as a sharpening stone adapted to your bird species.
The maximum size of the two birds should not exceed that of a budgerigar.

11. Feeding houses for birds in the winter

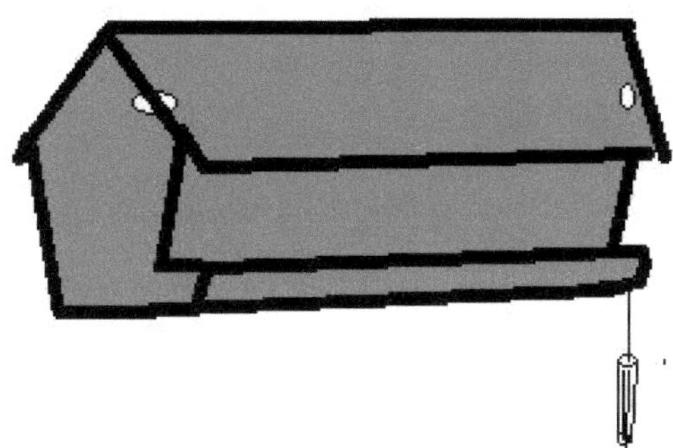

When snow falls in the winter, not all birds find enough food.
First of all, you have to consider whether you make it from plywood because of the small size or from boards. If you make it from plywood, you should replace nailing with gluing.
Feeding houses are hung so that a cat cannot catch the birds.
Most have a length of 30cm, 20cm height and 20cm width.

Materials Requirements
3x 30x20cm boards (plywood/wood)
2x 30x15cm pieces (roof)
2x 20cm² (side walls)
1x 30x 14.5cm (rear wall)
1x 30x12cm (front wall)
1x Bamboo, craft round wood 30cm long
2 slices of wood, plywood Ø 12cm
1 wooden or metal ring inner diameter 9cm (**stainless**)
6 bamboo sticks or wooden sticks Maximum Ø 1cm
1 approx. 25cm natural rope (rope for hanging the corn balls)
Flower wire
2 Hinges, 2 small metal angles, 4 screws for the angles and screws corresponding to the hinges, a small metal rod that fitting through the holes of the angles (old knitting needle …) 35cm long
>19 nails 5-6cm long

Tools: jigsaw, hammer (or brush for the wood glue), ruler 30cm, old craft needle, drill with a wood drill 1cm Ø, cordless screwdriver

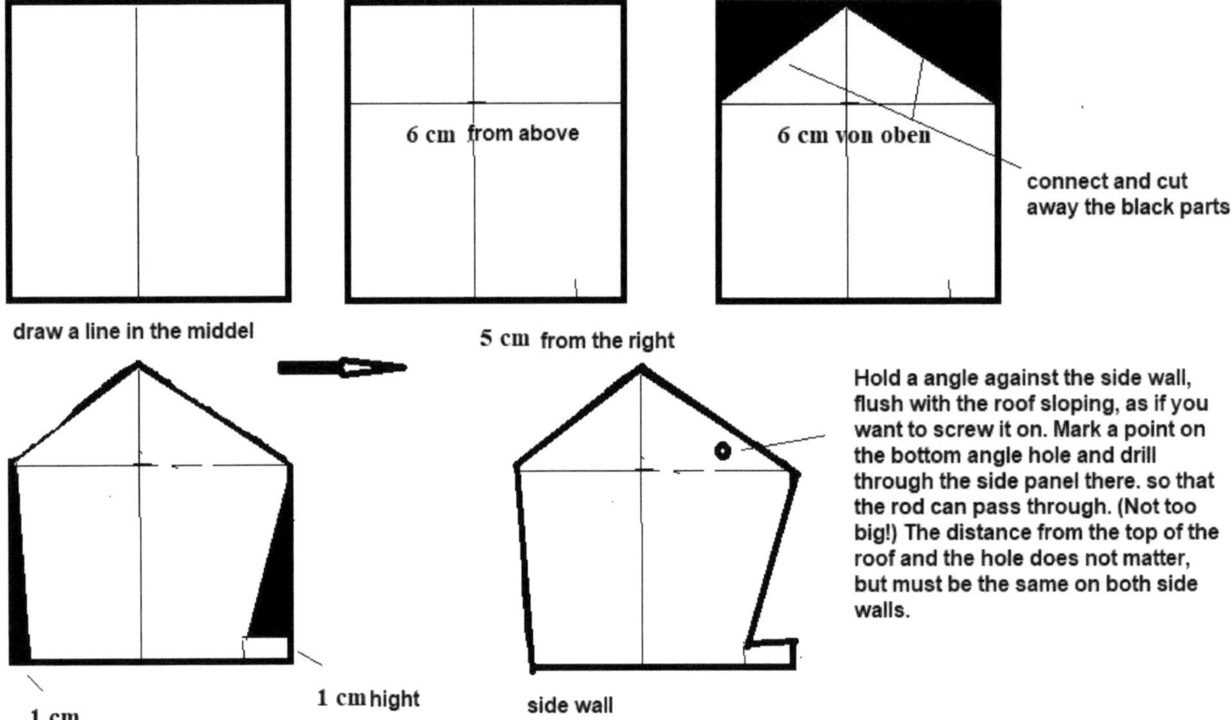

Production of the side walls:

Follow the picture instructions from right to left.

Now nail or glue the rear roof to the rear and side walls in a way that the roof
protrudes above, in the board thickness of the roof boards
Second, nail the ground.
(The floor should stick out a bit in front.)
Then nail the front wall, flush at the top, so that there is a slot at the bottom that
allows the lining to slide.
Now connect the other roof board with the nailed one through the hinges (hinges on
the outside). Mount the two angles on the inside of the front roof so that they can
open and close the front roof and fit the angle hole with the side walls for the rod.
You can close the front roof and push the rod through. Now the wind or clever birds
cannot open the roof and no snow or rain can enter.

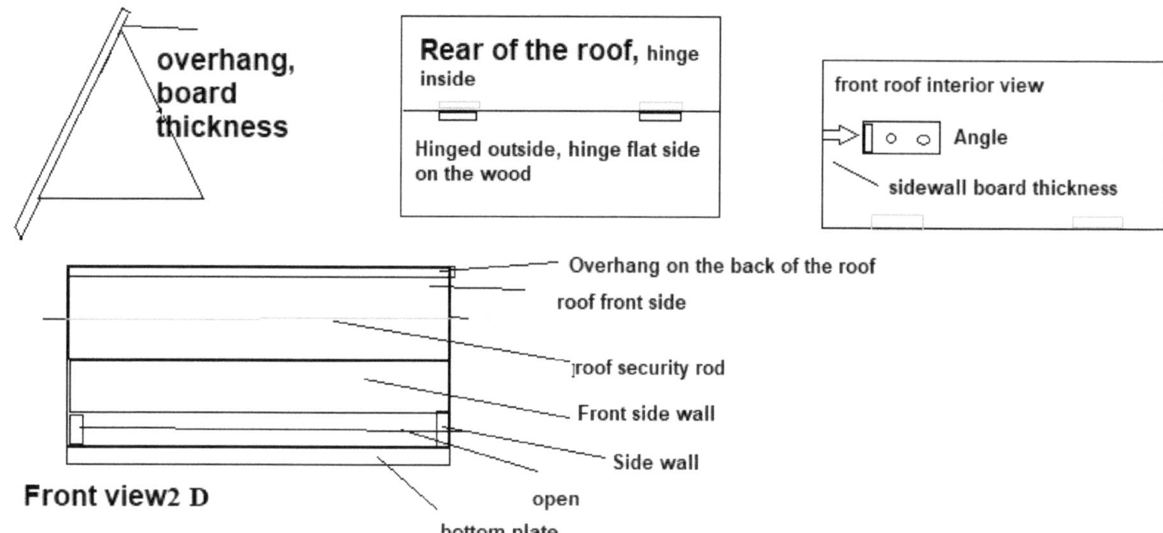

The corn balls have a diameter of about 8 cm.

Take a disc from the wooden discs and drill 6 holes at the edge.

Take the sticks and poke a hole at both ends so that a flower wire passes through. Put the rods through the holes of the disc so that on the other side you can thread the flower wire into the rods and knot into a ring. Pull back the sticks so that the flower wire is close to the disc.

Place the ring on the other rod ends. Wrap around the ring, the flower wire and pull the wire through the rods (see sketch).

Now drill through the other disk 2 holes, facing each other.

Thread the rope through and tie it to the ring. So that the upper disc functions as a cover that can be pulled up and down along the rope.

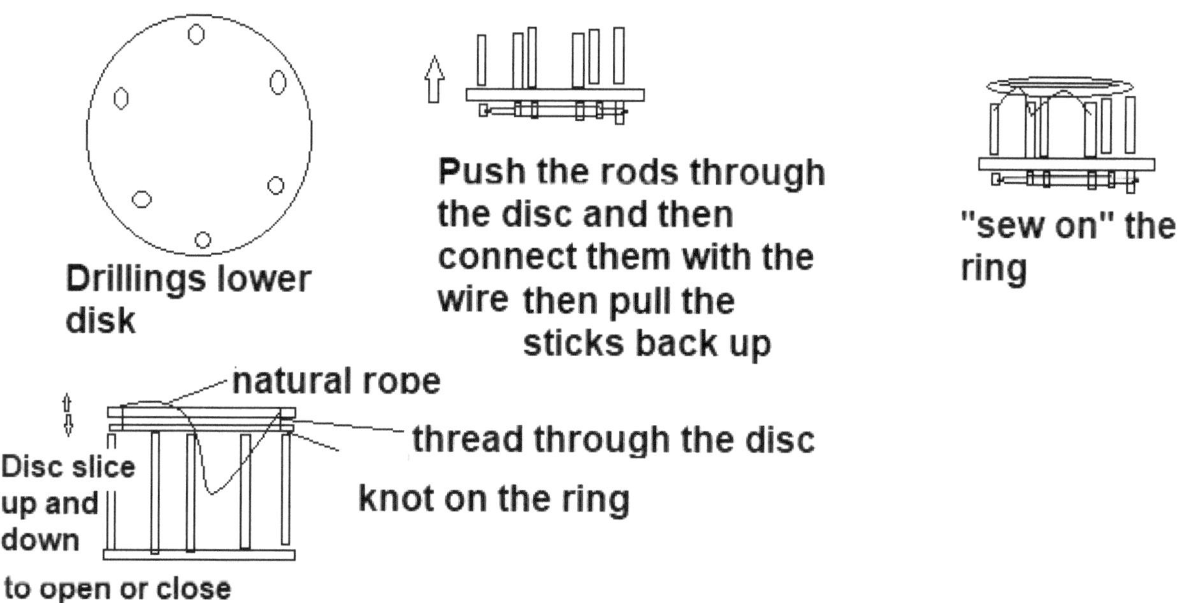

Hang the string around the 30cm long rod and nail it to the front bottom edge so that it lies in the corner of a side wall and bottom.

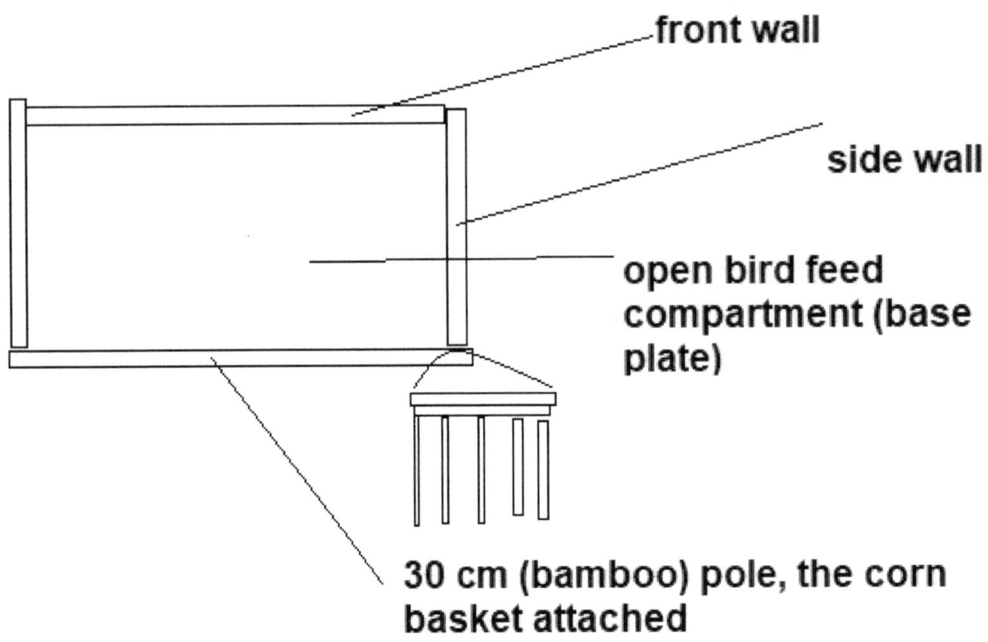

12. Nest boxes/houses

Nest boxes and cats only fit together when the nest boxes are attached out of reach of the cat, otherwise the nest boxes are hunting traps of the cats with the birds as prey. Theoretically, there are nest boxes, empty boxes with a hole in them, where birds pass through, although not always. If you nail a wooden box with boards and drill a hole in it for a blackbird to come in, it probably won't when the room is too big to keep warm. No, these birds do not go according to nature conservation law minimum place regulation, they decide during the apartment inspection themselves.
Sometimes nest boxes remain empty for years. Not because something is wrong with the boxes or because cats are nearby, but because it has not noticed that they can rent boxes by singing for you.

Here is a suggestion, you first build a feeder for the winter and /or simply observe, (before the feeder installation), which birds come to visit you.
Check the size of these birds on the internet or books.
The future mother must definitely fit through the hole. But only this bird species, because of egg robbery and the like.
There are nest boxes with and without landing pole in front of the opening.
Here I describe the construction of a 5cm nest box.

Requirement:
- 3 pieces of board 5 cm²
- 1 board 5 x 4cm (front)
- 1 board piece 5 cm² + 2x the board thickness wide (for example, you use boards that are 1cm thick then you need this floor board 5 x 7 cm, with a thickness of 2 cm the board must be 5 x 9cm…) as floor plate
- 1 board piece 9 – 10cm² as a roof
- 20 nails 25mm long (24 nails, if no hinge) or wood glue
- A small hinge and 4 size matching wood screws **[optional)**

There are people who say you have to clean the nest box in winter when everyone has flown out. However, if you build a nest box for birds that eat from your feed counter in the winter, a few will continue to inhabit the nest box.
In nature, no one makes the nest clean except the birds themselves.
If you use a hinge, you tightly close the roof for the nesting time.

Step 1:
Take 2 of the 5 cm² boards and mark them on one side at 1 cm and connect this mark to the opposite side so that they mark a small slope. (See image)
Take the small board and place it in front of you so that the height is 4cm.
Mark the middle for a hole. Drill a hole 12 mm in diameter.
The wooden drill bit is best suited for this purpose (see picture on page 9).
Continue to follow the drawing when nailing or gluing the parts together.
Take a nail at the side walls, top and center to nail them together. Then place the walls upside down and nail the floor flush to the walls. Then put the nest box back on the bottom side and nail the roof flush on the back wall at the walls.
When nailing again, make sure that NO nails look out with the tip. Alternatively, gluing with wood glue is also possible. In this case, the wood glue should not only be dried, but also ventilated before placing the nest box.
If you use a hinge, for example, and you do not clean the nest box, you can screw it on at the front with a wood screw, so that you can unscrew the screw for cleaning.

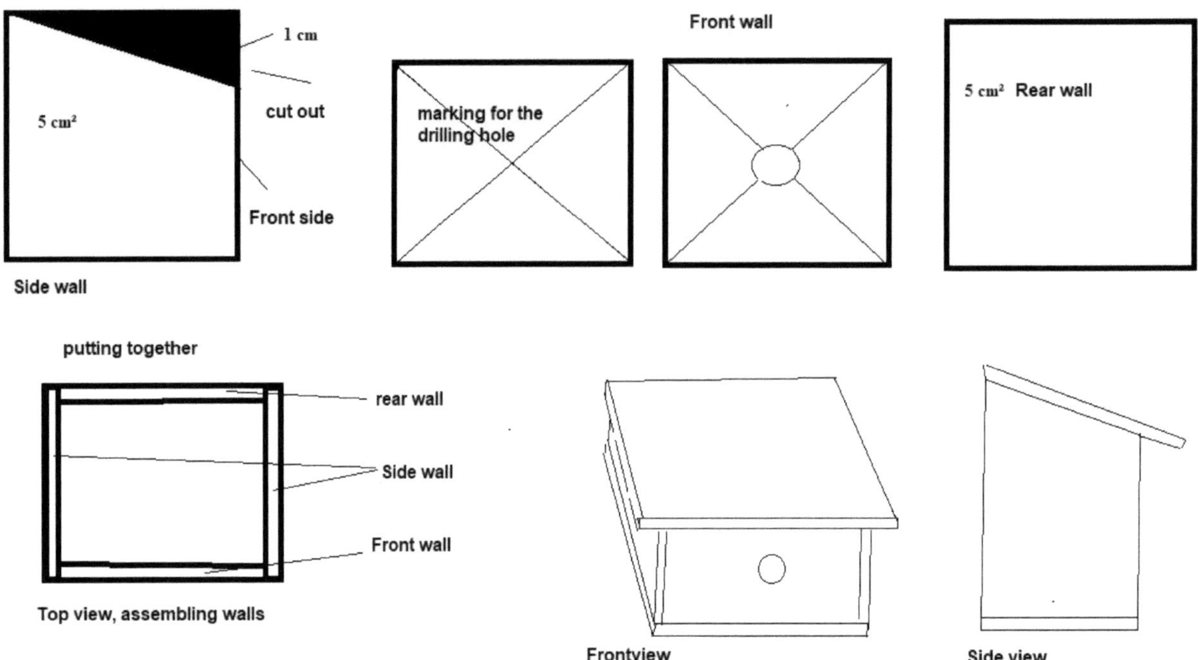

13. Insect hotel

As an insect hotel you can take the funniest shapes and simple construction methods. However, it is also important that a menu is served in front of the hotel. A super neat 1- 3 cm lawn is a zero menu for this. There are also fewer bees that inhabit an insect hotel. Rather beetles and if possible butterflies, which are caterpillars first. So at least there should be flowers nearby.

They don't have to be tulips. Cucumber, pumpkin, potato plant blossoms are also delicious. As well as cherries, apples or other fruit blossoms.

But if, for example, you sow or plant thyme and lavender around fruit trees, they keep ants and aphids away from the fruit trees but continue to have flowers when the fruit blossoms are already wrapped up in fruit.

It doesn't always have to be a bamboo tube.

Take a piece of squared wood and drill several holes in it.

For butterflies, you give wood wool, straw reed leaves cut in a free space in a plywood or craft board, slits that the butterflies pass through and glue or nail them in front of the free space with the wood wool.

Here is a small photo documentation. (Of course, you can cut all the squared timbers to one length beforehand to make it look more aesthetic. I have not done this because every gap can provide an insect shelter. For example, if a caterpillar wants to become a butterfly.)

 =>

 =>

I screwed the parts together with screws. Do not forget to make knots at the ends of the shear so that the cord does not untwist.
As a pile I used a small fence pile 1.5 meters with a peg.
This also allows me to relocate the hotel.

21 Construction and Animal Husbandry Act, as in Germany aplied

Chicken place requirement
36 cm² per chicken and/or rooster

Rabbits

For the keeping of breeding rabbits, this means:

1. The rabbit house must provide a floor area of at least 0.6 m² per rabbit
2. In addition, a nesting chamber of 0.1 m² must be available per animal
3. The ceiling height of the nest chamber must be at least 35 cm
4. The height from the top of the nest chamber to the top of the stable ceiling must also be at least 35 cm
5. There must be a floor with a floor area of at least 0.18 m²

For the keeping of fattening rabbits, this means:

1. The rabbit house must provide a floor area of at least 0.6 m² per rabbit

2.   If there is a nesting chamber, the height up to its ceiling must be at least 27 cm
3.   The distance from the top of the nest chamber to the stable ceiling must also be at least 27 cm
4.   The floor must be at least 0.15 m² in size

Recommendations of the Central Association of German Breed Rabbit Breeders e. V.

The "Central Association of German Rabbit Breeders e. V." (ZDRK e. V.) uses the following values in its guidebook for entering rabbit breeding (second edition).

Minimum dimensions for individual bays
*   For dwarf breeds under 1.5 kg: 50 cm high, 60 cm deep, 60 cm wide
*   For dwarf breeds over 1.5 kg: 50 cm high, 70 cm deep, 65 cm wide
*   For small breeds: 60 cm high, 75 cm deep, 70 cm wide
*   For medium sized breeds: 60 cm high, 80 cm deep, 85 cm wide
*   For large breeds: 70 cm high, 80 cm deep, 110 cm wide

Requirements for nest boxes and transport boxes: In this regard, the Central Association of German Breed Rabbit Breeders (Zentralverband Deutscher Rasse-Kaninchenzüchter e.V.) states the following values in its guide:

Nesting boxes
*   For dwarf breeds: 30 cm high, 30 cm deep, 30 cm wide
*   For small breeds: 35 cm high, 35 cm deep, 35 cm wide
*   For medium sized breeds: 40 cm high, 40 cm deep, 40 cm wide
*   For large breeds: 45 cm high, 60 cm deep, 45 cm wide

Transport boxes
*   For dwarf varieties: 25 cm high, 25 cm deep, 30 cm wide (area = 750 cm²)
*   For small breeds: 30 cm high, 25 cm deep, 35 cm wide (area = 875 cm²)
*   For medium sized breeds: 35 cm high, 30 cm deep, 45 cm wide (area = 1350 cm²)
*   For large breeds: 40 cm high, 35 cm deep, 55 cm wide (area = 1925 cm²)

As of 2023
<u>23 Project planning</u>

This project planning applies to the book "Garden animal dwellings made of wood" and the book coming up "Beautiful garden live in wood" and "Timber construction projects for your garden plants".
1. Read the complete guide once. For some, there are variants that make a difference in the material. Write down the optional additional materials and the general materials for your purchase.
2. Check your tools to see if they are still in order and if you have all the tools you need for construction.
3. Find a suitable place in the garden and measure whether this space is sufficient. Think of the stables also on the spout, and at the swings on the swivel range. Birds' nest boxes and/or feeders should not be easily reached by cats, but should be installed in such a way that you can clean or fill them. An insect hotel needs the sun to hatch eggs.

4. For larger projects such as the sitting area, chicken coop... you need space where you store the material and a weatherproof cover. Or if you can't buy the material at once, a space to collect the material.
5. Choose the most service-friendly hardware store – in your area.
    1. Can I order everything in advance?
    2. How much does it cost to deliver large parts (if you can't pick it up yourself).
    3. Do I get good advice for colorants and protective products if I show the consultant my list of materials there?
    4. Doe I get the nails and screws in small and larger quantities. (Don't buy exact quantities, a nail can get crooked...).
    5. Weatherproof pretreated materials such as boards and panels are usually not suitable for the inside of animal housing. Ask the dealer for the appropriate remedy.
6. After purchasing the materials, check that all the required parts are present.
7. Now pre-treat all the wooden parts and let them dry for the time indicated on the packaging.
8. Now read the construction instructions again completely and look at the sketches and then follow the steps of the instructions.
9. Once the construction is finished, you make a stability test, e.g. sit once on the roof. Everything is stable enough for the chickens. Sit on top of the doghouse. Or put yourself on the children's swing. Take the swing (hung) in your hand, arms stretched out all the way up and brace yourself with the full weight hanging on the swing against it. (Further stability tests in the instructions.)

Sources: https://en.wikipedia.org/wiki/Edelholz
https://www.bundestag.de/resource/blob/928356/57ef755d70en0771c565049627198380/WD-5-150-22-pdf-data.pdf
https://www.provieh.de/2016/04/gesetzliche-vorgaben-fuer-die-huehnerhaltung-im-eigenen-garten/
https://www.huehner-haltung.de/wissen/hintergrundwissen/gesetze/
https://kaninchenwiese.de/haltung/hintergruende/mindestmass-2/
https://www.kaninchen-haltung.com/kaninchenstall/vorschriften-kaninchenstall-und-kaninchenhaltung/
https://www.fuenftepfote.de/hundehuette-outdoor/hundehuette-groesse-berechnen/

The project instructions here are for people who prefer doing things by themselves rather than buying those things ready made. And perhaps someone who has built something from these instructions, will adapt some details according to his own taste. It perhaps promotes one or the other DIY store, because you buy your material there, but I do not see this book as a competition to industry.
Here, building and crafting must be fun. If you don't find fun in such a thing, you will also find all these things in the animal markets and construction houses.
But it is also partly so that those who have good money, but do not want to take the time, can have it made to measure, perhaps by you according to this book.

Notice !!! According to the information of the building office in my place of residence, in Germany the individual regulation for fixed buildings on a property is the responsibility of the local responsible building offices.

As a professional, I can, of course, also create an individual plan, which you can order from me at AutorRaginmund@gmail.com.

Other books by me

	Titel	Autor*in	Medium	ISBN	Veröffentlichung
	Broken Code english	Raginmund, -	Buch E-Book	9783743192249 9783757836184	03.04.2023
	Home Terra Preta - home made black soil english	Raginmund	Buch E-Book	9783756231966 9783756245437	29.06.2022
	Home Terra Preta - Hausmacher Schwarzerde	Raginmund	Buch E-Book	9783755734239 9783756245413	29.06.2022
	Dan's Adventure in Africa english	Raginmund	Buch E-Book	9783837059175 9783755794363	25.02.2022
	Dan's Abenteuer in Afrika	Raginmund	Buch E-Book	9783752816198 9783752817997	09.04.2018
	Broken Code	Raginmund	Buch E-Book	9783738608038 9783739290218	28.05.2015

Manufacturing and Publishing:
BoD - Books on Demand,
Norderstedt (Germany)

ISBN 9783757853525

© 2023 Raginmund all copy right infos by Sabam.be

MIX
Papier aus verantwortungsvollen Quellen
Paper from responsible sources
FSC® C105338